Easy Learning

GCSE Higher

science

Exam Practice Workbook

FOR OCR GATEWAY B

Contents

Fit for life

1 Ben goes running to get fit.

  a After a while Ben's muscles start to hurt.

    i Name the chemical made in Ben's muscles that causes the pain.

    _____ [1 mark]

    ii Explain why the muscles make this chemical.

    _____
    _____
    _____ [3 marks]

  b After the run Ben continues to breathe heavily and his heart rate stays high. Explain why.

    _____
    _____
    _____ [2 marks]

D–C

B–A*

2 Ben is an athlete and he is very fit. However, Ben still catches a cold. Explain why being fit may not keep you healthy.

_____
_____ [2 marks]

D–C

3 Amelia has her blood pressure checked by a nurse. The nurse tells Amelia her blood pressure is too high. She fills in a questionnaire for the nurse.

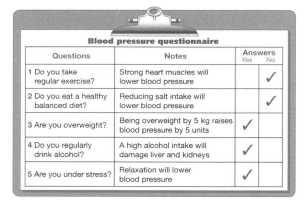

Blood pressure questionnaire

Questions	Notes	Answers Yes	Answers No
1 Do you take regular exercise?	Strong heart muscles will lower blood pressure		✓
2 Do you eat a healthy balanced diet?	Reducing salt intake will lower blood pressure		✓
3 Are you overweight?	Being overweight by 5 kg raises blood pressure by 5 units	✓	
4 Do you regularly drink alcohol?	A high alcohol intake will damage liver and kidneys	✓	
5 Are you under stress?	Relaxation will lower blood pressure	✓	

  a Suggest **two** changes Amelia should make to lower her blood pressure.

    1 _____
    _____
    2 _____
    _____ [2 marks]

  b Describe the possible consequences of high blood pressure for Amelia.

    _____
    _____ [2 marks]

D–C

B–A*

What's for lunch?

1 African children sometimes have a swollen abdomen, a condition caused by a low protein diet.

D–C

    a Write down the name of this condition.

    _____ [1 mark]

    b A child has a mass of 40 kg. Calculate his recommended average protein intake (RDA) in grams. Use this formula: RDA in g = 0.75 x body mass in kg
Show your working.

        RDA = _____ g   [2 marks]

2 Simon is overweight.

D–C

    a Simon would like to lose weight. He decides to change his diet. Suggest **one other** way Simon could lose weight.

    _____ [1 mark]

B–A*

    b Simon is a vegetarian. He needs to think carefully about his protein intake. Explain why.

    _____

    _____

    _____ [2 marks]

3 Enzymes are used to digest food.

D–C

    a Finish the table to name the enzyme that digests each type of food and the product of the digestion. The first one has been done for you.   [4 marks]

food type	enzyme	product
starch	*carbohydrase*	*glucose*
protein		
fat		

B–A*

    b Use ideas about **enzymes**, **bile** and **size of droplets** to explain how fats are digested in the body.

    _____

    _____

    _____

    _____ [3 marks]

Keeping healthy

1 Look at the diagram. It shows how mosquitoes spread malaria.

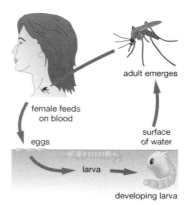

adult emerges

female feeds
on blood

surface
of water

eggs

larva → developing larva

a i What name is given to animals, such as the mosquito, that carry pathogens?

_____ [1 mark]

   ii Malaria is caused by a parasite called *plasmodium falciparum*.
      Explain why it is a parasite.

_____ [1 mark]

b Use the diagram to explain **one** way in which the spread of malaria could be controlled.

_____
_____ [2 marks]

D–C

B–A*

---

2 a Explain the difference between active and passive immunity.

_____
_____
_____ [3 marks]

b Vaccinations can contain harmless forms of a pathogen. Vaccinations can cause active immunity. Explain how.

_____
_____ [2 marks]

c Antibiotics are drugs used to treat some infections.

   i Explain why antibiotics cannot be used to treat a viral infection.

_____ [1 mark]

   ii Suggest a reason why doctors are concerned about the over-use of antibiotics.

_____ [1 mark]

   iii Doctors need to develop new drugs such as antibiotics. Describe how a blind trial can be used to test a new drug.

_____
_____
_____ [2 marks]

D–C

B–A*

D–C

B–A*

Keeping in touch

1 a Finish the table to show which part of the eye performs which function. The first one has been done for you.

[3 marks]

part of the eye	function
iris	*controls the amount of light entering the eye*
retina	
optic nerve	
cornea	

b The owl uses both eyes to see the same image.
Explain the advantage this type of vision gives the owl when it hunts.

_____

_____ [1 mark]

c Use ideas about **ciliary muscles**, **suspensory ligaments** and **lens** to describe how the eye focuses on a distant object.

_____

_____

_____ [3 marks]

d Some people find it difficult to focus on distance objects because they are short-sighted. Write down **two** different ways in which short sight can be corrected.

1 _____

2 _____ [2 marks]

2 The diagram shows the pathway taken by the impulse during an automatic reaction.

a Finish labelling the diagram to show the neurones involved in this process.

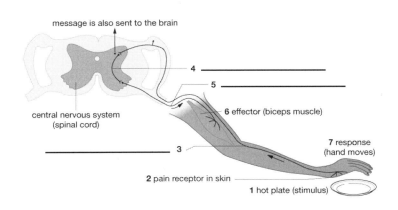

message is also sent to the brain

4 _____

5 _____

6 effector (biceps muscle)

central nervous system (spinal cord)

_____ 3

7 response (hand moves)

2 pain receptor in skin

1 hot plate (stimulus)

[3 marks]

b Describe how the impulse moves from one neurone to the next across the synapse.

_____

_____ [2 marks]

Drugs and you

1 a Different types of drugs have different effects. Finish the table by naming **one** example of each type of drug. Choose words from this list.

D–C

alcohol   anabolic steroid   cannabis   caffeine   heroin

type of drug	example
hallucinogen	
depressant	

[2 marks]

b Some people believe that personal use of cannabis should be allowed. Suggest **one** argument for and **one** argument against legal use of cannabis.

B–A*

For _____

_____

Against _____

_____

[2 marks]

2 a Describe the effects of tobacco smoke on epithelial cells in the lining of the trachea.

D–C

_____

_____

[2 marks]

b Describe the effect of nicotine on synapses.

B–A*

_____

_____

[2 marks]

3 a The picture shows drinks that contain one unit of alcohol. Matthew drinks two pints of beer and a glass of whisky. Jo drinks three glasses of wine and a glass of whisky. Who drinks the most units? Explain your answer.

D–C

half pint beer   single whisky   glass of wine

_____

_____

[2 marks]

b There are more alcohol-related injuries ocurring on a Saturday compared to a Wednesday night. Suggest why.

B–A*

_____

_____

_____

[2 marks]

Staying in balance

D–C

1 a The body can lose or gain heat. If the body gets too hot you can suffer from dehydration. Explain why.

_____

_____ [2 marks]

b When the body gets too cold the pulse rate slows. Name the condition the body suffers from when it gets too cold.

_____ [1 mark]

B–A*

c To prevent the body getting too warm vasodilation takes place. Explain what is meant by the term **vasodilation**.

_____

_____ [2 marks]

d Controlling body temperature involves negative feedback. Explain why.

_____

_____ [2 marks]

D–C

2 Blood sugar level is controlled by the hormone insulin.

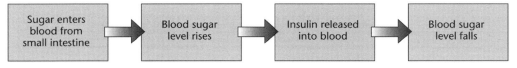

| Sugar enters blood from small intestine | → | Blood sugar level rises | → | Insulin released into blood | → | Blood sugar level falls |

a Some people do not make enough insulin. Write down the name of the condition they suffer from.

_____ [1 mark]

b Suggest **one** way such people can control their blood sugar level.

_____ [1 mark]

B–A*

3 This diagram shows the menstrual cycle.

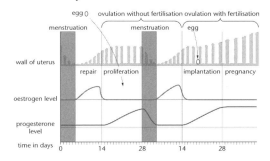

a Describe the effect of oestrogen on the lining of the uterus.

_____ [1 mark]

b Describe the effect of fertilisation on progesterone levels.

_____ [1 mark]

c Ovulation does not take place in some women. Suggest **one** way in which they can be treated.

_____ [1 mark]

Gene control

1 Finish the sentences about chromosomes.

Chromosomes are found in the _____ of the cell. They carry instructions called

_____. The chromosomes are made of a chemical called _____.   [3 marks]

D–C

2 The squirrel developed from a fertilised egg. The fertilised egg is made up of 20 chromosomes.

D–C

   a How many chromosomes are in the egg before it is fertilised by a sperm?

      _____   [1 mark]

   b How many chromosomes are in one cell from the squirrel's ear?

      _____   [1 mark]

   c The nucleus of a human sperm is different from the nucleus of a squirrel sperm. Describe how they are different.

      _____   [1 mark]

   d The cells in the squirrel's ears contain a code to make insulin but only the squirrel's liver cells make insulin. Explain why.

B–A*

      _____

      _____   [2 marks]

3 Chromosomes contain special chemicals called bases.

   a How many different bases are there in a chromosome?

D–C

      _____   [1 mark]

   b If a section of chromosome contained 30 **T** bases how many **A** bases will it contain? Explain your answer.

B–A*

      _____

      _____   [2 marks]

   c This diagram shows how DNA controls eye colour. Changing the DNA code would change the eye colour. Explain how.

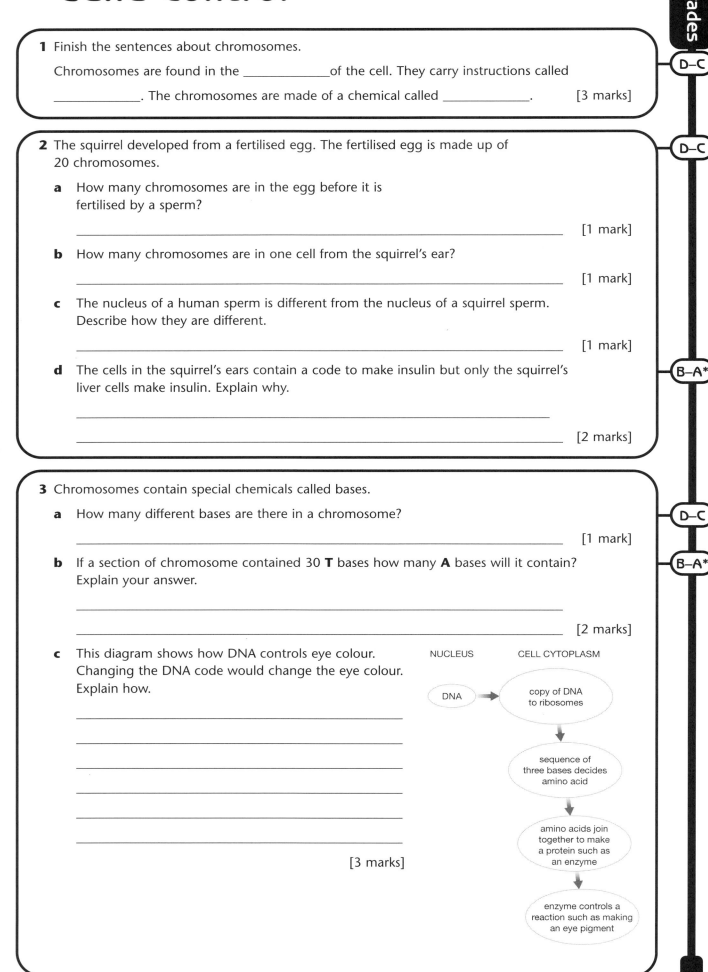

NUCLEUS   CELL CYTOPLASM

DNA → copy of DNA to ribosomes

sequence of three bases decides amino acid

amino acids join together to make a protein such as an enzyme

enzyme controls a reaction such as making an eye pigment

      _____

      _____

      _____

      _____

      _____

      _____

      [3 marks]

Who am I?

1 We inherit different characteristics from our parents' chromosomes.

D–C

a Finish the table to show how gender is inherited. [4 marks]

egg	sperm	fertilised egg	gender of child
X	X		
X	Y		

B–A*

b The sperm determine gender. Use your knowledge of genetics to explain this.

_____

_____

_____ [2 marks]

2 Mutations are changes to genes. Mutations can occur spontaneously or they can be caused by some other factors.

D–C

a Write down **one** factor that can cause mutations.

_____ [1 mark]

B–A*

b Haemophilia is a condition caused by a gene mutation. A protein needed to clot the blood cannot be made by the body. Explain why the mutation stops the protein being made.

_____

_____ [2 marks]

D–C

3 a The diagram shows a breeding experiment.

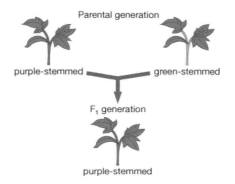

Parental generation

purple-stemmed ——— green-stemmed

F_1 generation

purple-stemmed

Which characteristic is dominant? Explain your answer.

_____ [1 mark]

B–A*

b Cystic fibrosis is an inherited condition. Sam has cystic fibrosis but his parents are normal.

i Finish the genetic diagram to show how Sam inherited cystic fibrosis.

ii Put a (ring) around Sam's genotype on the diagram.

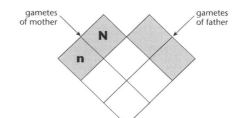

gametes of mother

N

n

gametes of father

[2 marks]

B1 Revision checklist

- I can explain the possible consequences of having high blood pressure. ☐

- I can state the word equation for respiration. ☐

- I know that religion, personal choice and medical issues can influence diet. ☐

- I can calculate BMI and RDA. ☐

- I can describe how white blood cells defend against pathogens. ☐

- I can explain the difference between passive and active immunity. ☐

- I can describe the functions of the main parts of the eye. ☐

- I know the classification and effects of the different types of drugs. ☐

- I can interpret data on the effects of smoking. ☐

- I can describe vasodilation and vasoconstriction. ☐

- I know that sex hormones can be used as contraceptives and for fertility treatment. ☐

- I know that chromosomes are made of DNA. ☐

- I can carry out a genetic cross to predict the possibilities of inherited disorders passing to offspring. ☐

- I know that inherited characteristics can be dominant or recessive. ☐

Cooking

D–C

1 a Some foods need to be cooked. Explain why.

_____

_____ [2 marks]

D–C

2 a Write down **two** good sources of carbohydrate.

_____ [2 marks]

b Write down **two** good sources of protein.

_____ [2 marks]

B–A*

c Protein molecules change when they are cooked. Explain why this is important.

_____

_____

_____ [3 marks]

D–C

3 a Baking powder is a chemical called **sodium hydrogencarbonate**. When it is heated it **decomposes** to give sodium carbonate, carbon dioxide and water.

i Write down the word equation for the reaction.

_____

_____ [2 marks]

ii Write down the **reactants** of the reaction.

_____ [1 mark]

iii Write down the **products** of the reaction.

_____ [1 mark]

B–A*

b The formula for sodium hydrogencarbonate is $NaHCO_3$. Write down the balanced symbol equation for this reaction.

_____

_____ [3 marks]

D–C

4 The chemical test for carbon dioxide is to pass it through limewater. It will turn the

limewater from _____

to _____ .

carbon dioxide ⟶

delivery tube

limewater

[2 marks]

Food additives

1 a Why are antioxidants added to tinned fruit?

_____ [1 mark]

b Ascorbic acid (vitamin C) is used as an antioxidant in which foods?

_____

_____ [2 marks]

c Information about food is given on food labels. How much energy does this food provide?

Energy	56 J
Sodium	0.02 mg
Magnesium	Trace

_____ [1 mark]

D–C

2 a Why is food packaging used?

_____ [1 mark]

b What is active packaging?

_____ [1 mark]

c Why is active packaging beginning to be used?

_____ [1 mark]

d It uses a **polymer**, what else is needed to make it work?

_____ [1 mark]

e Intelligent packaging works by including **indicators** on packages. Explain how these work.

_____ [1 mark]

D–C

B–A*

3 a Look at the diagram.
It is a detergent molecules made up of two parts, a head and a tail. Describe how the detergent works on removing grease in water. Use the diagram to help you.

fat-loving part

water-loving part

grease spot

_____
_____
_____
_____ [5 marks]

b Mayonnaise is made of oil, vinegar and egg yolk. This makes a smooth substance. Explain how. Use the words **hydrophilic** and **hydrophobic** in your answer. Use a diagram to help you.

_____
_____
_____
_____ [4 marks]

D–C

B–A*

Smells

D–C

1 a To make a perfume alcohol is mixed with an acid
to make an ester.

  i Write down a word equation for this reaction.

  _____ [2 marks]

  ii Look at the diagram. Label the alcohol and acid. [1 mark]
  iii Label the condenser. [1 mark]
  iv What is happening at **X**?

  _____
  _____ [1 mark]

  v Why is the condenser used?

  _____
  _____
  _____ [1 mark]

X —

D–C

2 A good perfume needs to have several properties. These are listed in the boxes.
Draw a **straight** line to match the **best** reason to the property needed.

evaporates easily		it can be put directly on the skin
non-toxic		its particles can reach the nose
insoluble in water		it does not poison people
does not irritate the skin		it cannot be washed off easily	[3 marks]

D–C

3 a A solute and a solvent that do not separate is a _____ . [1 mark]

  b Esters are used as perfumes and _____ . [1 mark]

B–A*

4 a If a liquid evaporates easily then the substance is volatile. Explain how this happens.
Use ideas about forces between particles in your answer.

  _____
  _____
  _____
  _____ [4 marks]

  b Water does not dissolve nail varnish. Use the
  diagram to help explain why.
  Use ideas about the forces of attraction between
  molecules in your answer.

  nail varnish
  molecule

  water molecules

  _____
  _____
  _____
  _____ [2 marks]

Making crude oil useful

1 a All the oils of crude oil are **hydrocarbons**. What is a hydrocarbon?

_____

_____
                                                                    [2 marks]

b The hydrocarbons are separated by **fractional distillation**.

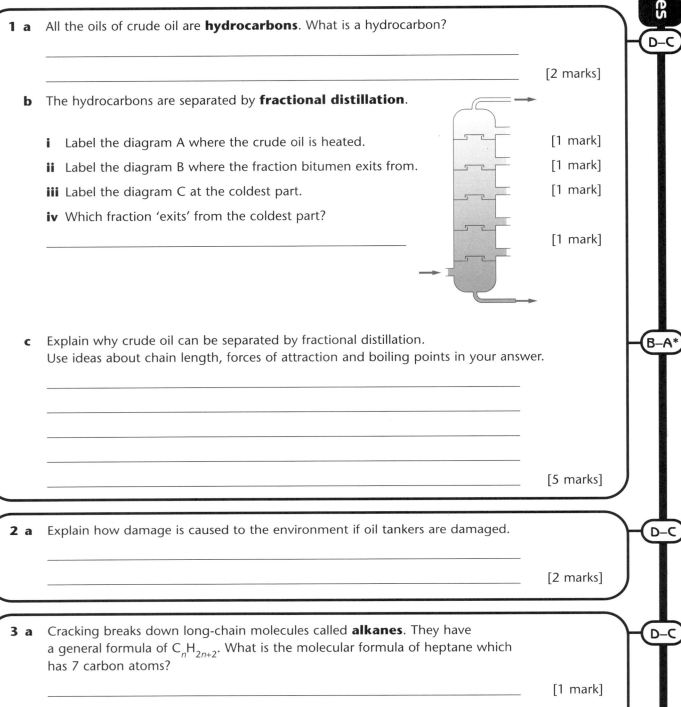

  i Label the diagram A where the crude oil is heated.   [1 mark]

  ii Label the diagram B where the fraction bitumen exits from. [1 mark]

  iii Label the diagram C at the coldest part.   [1 mark]

  iv Which fraction 'exits' from the coldest part?

_____   [1 mark]

c Explain why crude oil can be separated by fractional distillation.
Use ideas about chain length, forces of attraction and boiling points in your answer.

_____

_____

_____

_____

_____
                                                                    [5 marks]

D–C

B–A*

2 a Explain how damage is caused to the environment if oil tankers are damaged.

_____

_____
                                                                    [2 marks]

D–C

3 a Cracking breaks down long-chain molecules called **alkanes**. They have
a general formula of C_nH_{2n+2}. What is the molecular formula of heptane which
has 7 carbon atoms?

_____   [1 mark]

b When a large alkane is cracked it becomes a smaller alkane and an alkene.
Explain why an **alkene** is a different type of hydrocarbon to an alkane.

_____   [1 mark]

c What are alkenes useful for making? _____   [1 mark]

d A country produces 25% more than its demand of heavy oil from crude oil
distillation. However, its supply of petrol from the distillation is only 68% of its
need. Suggest how they could solve this problem other than by importing
petrol from elsewhere.

_____

_____

_____
                                                                    [3 marks]

D–C

B–A*

Making polymers

D–C

1 a Which molecule is a polymer? Put a (ring) around **A**, **B**, **C** or **D**.

A

$$H-\underset{\underset{H}{|}}{\overset{\overset{H}{|}}{C}}-\underset{\underset{H}{|}}{\overset{\overset{H}{|}}{C}}-OH$$

B

$$H-\underset{\underset{H}{|}}{\overset{\overset{H}{|}}{C}}-\underset{\underset{H}{|}}{\overset{\overset{H}{|}}{C}}-\underset{\underset{H}{|}}{\overset{\overset{H}{|}}{C}}-H$$

C

$$\left[\underset{\underset{H}{|}}{\overset{\overset{H}{|}}{C}}-\underset{\underset{H}{|}}{\overset{\overset{H}{|}}{C}}-\underset{\underset{H}{|}}{\overset{\overset{H}{|}}{C}}-\underset{\underset{H}{|}}{\overset{\overset{H}{|}}{C}}-\underset{\underset{H}{|}}{\overset{\overset{H}{|}}{C}}-\underset{\underset{H}{|}}{\overset{\overset{H}{|}}{C}}\right]_n$$

D

$$C-\underset{\underset{}{}}{\overset{\overset{H}{|}}{C}}=\underset{\underset{}{}}{\overset{\overset{H}{|}}{C}}-Cl$$

[1 mark]

b Write down **two** conditions needed for polymerisation.

_____

_____ [2 marks]

B–A*

c What does a monomer need in its structure to undergo addition polymerisation?

_____ [1 mark]

d Construct the **displayed formula** of the addition polymer from this monomer. [2 marks]

$$\underset{\underset{Cl}{|}}{\overset{\overset{H}{|}}{C}}=\underset{\underset{H}{|}}{\overset{\overset{H}{|}}{C}}$$

e Draw the monomer that makes this addition polymer. [2 marks]

$$\left[\underset{\underset{H}{|}}{\overset{\overset{H}{|}}{C}}-\underset{\underset{H}{|}}{\overset{\overset{CH_3}{|}}{C}}-\underset{\underset{H}{|}}{\overset{\overset{H}{|}}{C}}-\underset{\underset{H}{|}}{\overset{\overset{CH_3}{|}}{C}}-\underset{\underset{H}{|}}{\overset{\overset{H}{|}}{C}}-\underset{\underset{H}{|}}{\overset{\overset{CH_3}{|}}{C}}\right]_n$$

f Explain how addition polymerisation takes place.

_____

_____

_____

_____

_____ [4 marks]

D–C

2

a i Butanol, C_4H_9OH $H-\underset{\underset{H}{|}}{\overset{\overset{H}{|}}{C}}-\underset{\underset{H}{|}}{\overset{\overset{H}{|}}{C}}-\underset{\underset{H}{|}}{\overset{\overset{H}{|}}{C}}-\underset{\underset{H}{|}}{\overset{\overset{H}{|}}{C}}-OH$ is **not** a hydrocarbon. Explain why.

_____ [1 mark]

ii Butene is an alkene. $H-\underset{\underset{H}{|}}{\overset{\overset{H}{|}}{C}}-\underset{\underset{H}{|}}{\overset{\overset{H}{|}}{C}}-\underset{\underset{}{}}{\overset{\overset{H}{|}}{C}}=\underset{\underset{}{}}{\overset{\overset{H}{|}}{C}}-H$

Explain how you know.

_____ [1 mark]

B–A*

iii Butene is a **monomer**. What is **polybutene**?

_____ [1 mark]

b i Bromine solution is used to test for unsaturation. It is an orange solution. When an alkene is added the orange solution turns colourless. Explain why.

_____ [1 mark]

ii When bromine solution is added to an alkane, what do you see?

_____ [1 mark]

Designer polymers

1 a Polymers are better than other materials for some uses. Give **one** example of the use of a polymer in healthcare that is better than the material used before.

_____

_____ [2 marks]

b i Gore-Tex® type materials are used to make clothing waterproof and breathable. The inner layer of the clothing is made from expanded PTFE (polytetrafluoroethene) which is **hydrophobic**. Explain what hydrophobic means.

_____

_____ [2 marks]

ii The PTFE is expanded to form a **microporous membrane**. Explain how this makes a breathable membrane.

_____

_____ [2 marks]

2 a Scientists are developing addition polymers that are **biodegradable**. Explain why.

_____

_____ [2 marks]

b Suggest a use for a biodegradable plastic.

_____ [1 mark]

c Disposing of non-biodegradable polymers causes problems. Explain the problems for each of the ways of disposing.

Landfill sites _____

Burning waste plastic _____

Recycling _____ [3 marks]

3 Look at the diagram of polymer chains.

a i Label the **intramolecular covalent** bonds. [1 mark]

ii Label the **intermolecular** forces of attraction. [1 mark]

b Some plastics have low melting points and can be stretched easily whereas other plastics have high melting points and cannot be stretched easily. They are rigid. Explain why. Use ideas about forces between chains in your answer. You may use a diagram to help you.

_____

_____

_____

_____ [4 marks]

Using carbon fuels

D–C

1 a Look at the table. Which fuel produces more acid fumes?

characteristic	coal	petrol
energy value	high	high
availability	good	good
storage	bulky and dirty	volatile
toxicity	produces acid fumes	produces less acid fumes
pollution caused	acid rain, carbon dioxide and soot	carbon dioxide, nitrous oxides
ease of use	easy in power stations	easy in engines

_____ [1 mark]

ii Give **two** advantages of using either coal or petrol for heating.

_____

_____ [2 marks]

B–A*

b Explain why the use of petrol and diesel in transport is contributing to a global problem.

_____

_____

_____

_____ [4 marks]

D–C

2 a Write down a **word equation** for a hydrocarbon fuel burning in air.

_____ [1 mark]

b Two products are made in the complete combustion of a fuel. These can be tested in the laboratory.

i Limewater is used to test for one product. Which one?

_____ [1 mark]

ii How is the other product tested?

_____ [1 mark]

c Complete combustion is better than incomplete combustion. Explain why.

_____

_____

_____ [3 marks]

d Why should a room be well ventilated and a heater regularly checked?

_____

_____ [2 marks]

B–A*

e Write a balanced equation for the complete combustion of propane, C_3H_8.

_____ [2 marks]

Energy

1 a Use the words **exothermic** and **endothermic** correctly in these sentences.

When energy is transferred **out** to the surroundings in a chemical reaction it is an

_____ reaction (energy is released).

When energy is taken in from the surroundings in a chemical reaction it is an

_____ reaction (absorbs energy).

An_____ reaction is shown by a temperature **increase**.

Burning magnesium is an example of an _____ reaction.   [4 marks]

b Is bond breaking an exothermic or endothermic process?

_____   [1 mark]

c Burning methane is an exothermic reaction. Explain why, using ideas about bond breaking and bond making.

_____

_____

_____   [3 marks]

2 a The flame of a Bunsen burner changes colour depending on the amount of

oxygen that it burns in. The flame is _____ when the gas burns in plenty

of oxygen. This is because there is _____ combustion. The flame

is _____ when the gas burns in limited oxygen. This is because

there is _____ combustion.   [4 marks]

3 a Describe how you would design your own experiment to compare the energy transferred by two different fuels, using a beaker of water and a thermometer. Write down the ways to make the tests fair.

_____

_____

_____

_____

_____   [5 marks]

b Todd and Terri calculate the amount of energy transferred during a reaction. They use a spirit burner to burn the fuel to heat 100 g of water in a copper **calorimeter**. They measure a temperature change in the water of 50 °C. The mass of fuel they burn is 4.00 g.

i How do they make the tests reliable?

_____   [1 mark]

ii Show how they calculate the energy released per gram.
(The specific heat capacity of water is 4.2 J/g/°C)

[5 marks]

C1 Revision checklist

- I know that cooking food is a chemical change as a new substance is made and it is an irreversible reaction. ☐

- I know that protein molecules in eggs and meat change shape when the food is cooked. ☐

- I know that when the shape of a protein changes it is called denaturing. ☐

- I know that emulsifiers are molecules that have a water-loving part and an oil- or fat-loving part. ☐

- I know that alcohols react with acids to make an ester and water. ☐

- I know that a solute is the substance dissolved in a solvent to make a solution that does not separate. ☐

- I know that crude oil is a non-renewable fossil fuel, which is a mixture of many hydrocarbons. ☐

- I know that petrol is a crude oil fraction with a low boiling point, which exits at the top of the fractional distillation tower. ☐

- I know that polymerisation is a process which requires high pressure and a catalyst. ☐

- I know that a hydrocarbon is a compound formed between carbon atoms and hydrogen atoms only. ☐

- I know that alkenes are hydrocarbons with one or more double bonds between carbon atoms. ☐

- I know that complete combustion of a hydrocarbon fuel makes carbon dioxide and water only. ☐

- I know that an exothermic reaction is one where energy is released into the surroundings. ☐

- I can work out that: energy transferred = mass of water × 4.2 × temperature change. ☐

Heating houses

1 Kelly opens the front door on a very cold morning. Her mother complains that the house is getting cold.
Use your ideas about energy flow to explain why the house gets cold.

_____

_____ [2 marks]

D–C

2 A police helicopter uses a thermal imaging camera to take a picture at night. It is looking for a car that has recently been abandoned in a field after a high speed chase. How does the thermogram help locate the car?

_____

_____

_____ [3 marks]

B–A*

3 a Finish the sentence.
The energy needed to raise the temperature of 1 kg of a material by 1 °C is known as the _____. [1 mark]

D–C

b The specific heat capacity of seawater is 3900 J/kg °C. How much energy is needed to heat 500 g of seawater from 20 °C to 90 °C?
Show how you work out your answer.

_____ [4 marks]

B–A*

4 a What physical quantity is measured in units of J/kg?

_____ [1 mark]

D–C

b When iron changes from solid to liquid, energy is transferred but there is no temperature change until all of the solid has changed into liquid. Use your ideas about molecular structure to explain why there is no change in temperature at iron's melting point.

_____

_____

_____ [2 marks]

B–A*

Keeping homes warm

1 Dan wants to reduce the heating bills for his old house. He decides to insulate his loft first and then replace his windows with double glazing.

a Dan spends £120 on loft insulation. He is told that this will reduce his heating bills by £40 per year. Calculate the payback time for loft insulation.
Show how you work out your answer.

_____ [2 marks]

b Why does Dan decide to insulate the loft before replacing his windows?

_____

_____ [1 mark]

c Dan heats his house with coal fires. He is told that his fires are **32% efficient**. Explain what is meant by 32% efficient.

_____

_____ [2 marks]

d Dan pays £6.50 for a 25 kg bag of coal. How much of that money is **usefully** used in heating his house?

_____ [2 marks]

e Why are coal fires so inefficient?

_____

_____ [1 mark]

f What types of materials transfer energy?

i by conduction

_____ [1 mark]

ii by convection

_____ [1 mark]

iii by radiation

_____ [1 mark]

How insulation works

1 a The diagram shows a section through a double-glazed window. Michael says that it is just as effective to use a piece of glass twice the thickness. Use your ideas about energy transfer to explain why double glazing is better.

space filled with air or argon, or has a vacuum

[3 marks]

b New homes are built with insulation blocks in the cavity between the inner and outer walls. The blocks have shiny foil on both sides.

block wall

exterior brick or stone finish

solid foam board

i Explain how the insulation blocks reduce energy transfer by conduction and convection.

[4 marks]

ii Explain how the shiny foil helps to keep a home warmer in winter and cooler in summer.

[2 marks]

2 a A brick in a wall is a better conductor of heat than air in the cavity between the walls.
Use your ideas about particles and kinetic energy to explain this.

[3 marks]

b Hot air rises. Explain why.

[3 marks]

Cooking with waves

1 a Why do microwave ovens take less time to cook food than normal ovens?

_____ [1 mark]

b Microwaves are suitable to communicate with spacecraft thousands of kilometres away while sometimes a mobile phone cannot receive a signal just a few kilometres from the nearest transmitter. Why do microwave signals seem to work better in space than they do on Earth?

_____

_____ [2 marks]

c Dave says that microwave signals 'bounce' off satellites. Sally says Dave is wrong. What happens to a microwave signal when it is received by a satellite?

_____

_____ [2 marks]

2 The diagram shows the electromagnetic spectrum.

← increasing wavelength   increasing energy →

wavelength

visible light

radio waves   infrared   ultraviolet   gamma rays

microwaves   X - rays

a Which part of the electromagnetic spectrum transfers the most energy? _____

[1 mark]

b An electric iron transfers less energy than the element of an electric fire. How does the wavelength of radiation from the iron differ from the wavelength of radiation from the element?

_____

_____ [1 mark]

3 The diagram shows a transmitter on top of a hill. It transmits microwave mobile phone signals as well as radio and television signals.

The house, behind the other hill, can receive radio and television signals but there is no mobile phone reception. Use your knowledge of microwave properties to explain why.

_____

_____

_____ [2 marks]

Infrared signals

1 a The diagram shows a signal displayed on an oscilloscope. What type of signal is it? Put a (ring) around the correct answer.

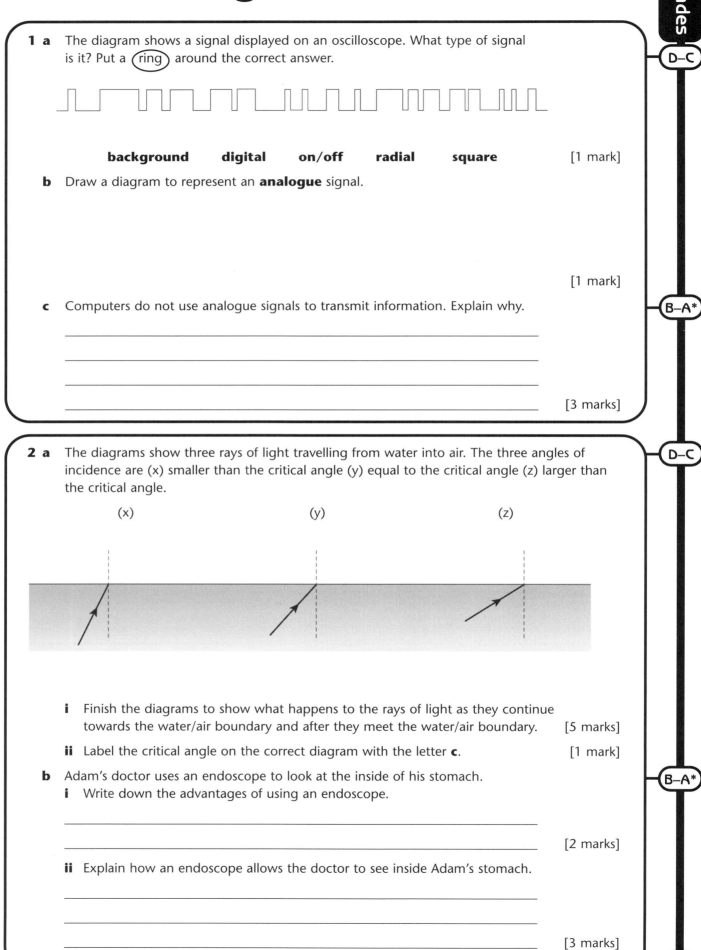

      background   **digital**   **on/off**   **radial**   **square**   [1 mark]

b Draw a diagram to represent an **analogue** signal.

[1 mark]

c Computers do not use analogue signals to transmit information. Explain why.

_____

_____

_____

_____ [3 marks]

2 a The diagrams show three rays of light travelling from water into air. The three angles of incidence are (x) smaller than the critical angle (y) equal to the critical angle (z) larger than the critical angle.

        (x)   (y)   (z)

    i Finish the diagrams to show what happens to the rays of light as they continue towards the water/air boundary and after they meet the water/air boundary. [5 marks]

    ii Label the critical angle on the correct diagram with the letter **c**. [1 mark]

b Adam's doctor uses an endoscope to look at the inside of his stomach.
    i Write down the advantages of using an endoscope.

_____

_____ [2 marks]

    ii Explain how an endoscope allows the doctor to see inside Adam's stomach.

_____

_____

_____ [3 marks]

Wireless signals

D–C

1 Radio waves are refracted in the upper atmosphere. How is the amount of refraction affected if the frequency of the radio wave is decreased?

_____

[1 mark]

B–A*

2 Luke and Jessica are walking in the mountains. They are both listening to their personal radios.

Luke is listening to Radio 1 on FM and Jessica is listening to Radio 4 on long wave. Both radio stations are broadcasting from the same aerial. Luke complains that he keeps losing the signal but Jessica does not have this problem.

Use your ideas about waves to explain why Luke has problems with reception but Jessica does not.

_____

_____

_____

[3 marks]

B–A*

3 A satellite is in orbit, 36 000 km above the equator. It takes 24 hours to orbit Earth. Microwaves travel at the speed of light (300 000 km/s).

a Why does a communications satellite take 24 hours to orbit Earth?

_____

[1 mark]

b How long does it takes for a microwave signal to travel from Earth, to the satellite and back to Earth? Put a (ring) around the correct answer.

0.12 s   0.24 s   1.2 s   2.4 s

[1 mark]

D–C

4 a Ella listens to her favourite radio station. Every so often, she notices that she can hear a foreign radio station as well. Put ticks (✓) in the boxes next to the **two** statements that explain why this happens.

The foreign radio station is broadcasting on the same frequency.	
The foreign radio station is broadcasting with a more powerful transmitter.	
The radio waves travel further because of weather conditions.	
Ella's radio needs new batteries.	

[2 marks]

B–A*

b Why does the microwave beam sent by a transmitting aerial towards a satellite in orbit have to be **focused**?

_____

[1 mark]

Light

1 The diagram shows a transverse wave.

D–C

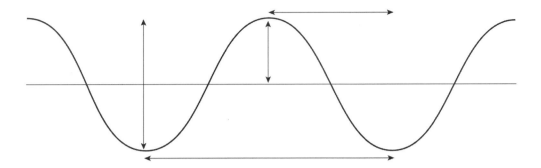

a Write the letter **A** next to the arrow which shows the amplitude of the wave. [1 mark]

b Write the letter **W** next to the arrow which shows the wavelength of the wave. [1 mark]

c What is meant by the **frequency** of a wave?

_____

_____ [1 mark]

2 a Why was it necessary for Samuel Morse to devise a code of dots and dashes?

D–C

_____ [1 mark]

b Lisa is using a signalling lamp to send a message to Robbie. Write down **one** advantage and **one** disadvantage of using a signalling lamp.

B–A*

Advantage _____

_____

Disadvantage _____

_____ [2 marks]

c Lisa's science teacher shines a laser light onto a screen. It is brighter than the white light from the laboratory lights. Use your ideas about frequency to explain the difference between laser light and white light. Draw diagrams to help you answer the question.

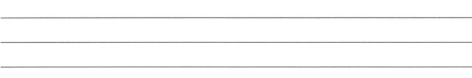

_____

_____

_____

[4 marks]

Stable Earth

D–C

1 a **P** waves and **S** waves are two of the waves which travel through the Earth after an earthquake. Put a tick (✓) in the box or boxes to correctly describe each wave. The first one has been done for you.

description	P wave	S wave
pressure wave	✓	
transverse wave		
longitudinal wave		
travels through solid		
travels through liquid		

[5 marks]

B–A*

b How do scientists use the properties of **P** waves and **S** waves to find out the **size** of the Earth's core?

_____

_____ [2 marks]

c How do scientists use the properties of **P** waves and **S** waves to find out the **structure** of the Earth's core?

_____

_____ [2 marks]

D–C

2 a Why does dust from factory chimneys cause warming of the Earth?

_____ [1 mark]

B–A*

b What effect do CFCs from aerosols and refrigerators have on the ozone layer?

_____

_____ [2 marks]

c Why is it beneficial to Earth to have a layer of ozone surrounding the planet?

_____ [1 mark]

D–C

d Leah wants to sunbathe and get a good tan.

i What causes Leah's skin to tan?

_____

_____ [2 marks]

ii She uses a sun screen with **SPF 30**. What does SPF 30 mean?

_____

_____ [2 marks]

P1 Revision checklist

- I can explain the difference between temperature and heat. ☐

- I can explain what is meant by specific heat capacity and specific latent heat. ☐

- I can perform energy calculations involving specific heat capacity and specific latent heat. ☐

- I can explain why different forms of domestic insulation work in terms of energy transfer. ☐

- I can calculate energy efficiency. ☐

- I can explain why infrared radiation is used for cooking. ☐

- I can explain why microwaves are used for cooking and for communication. ☐

- I can explain the advantages of using digital signals instead of analogue signals. ☐

- I can describe total internal reflection and its use in optical fibres, including the endoscope. ☐

- I can explain how wireless signals are used for global communication. ☐

- I can explain why there is sometimes interference with microwave and radio signals. ☐

- I know the properties of a transverse wave and how light is used as a transverse wave. ☐

- I can describe earthquake waves and how they can tell us about the structure of the Earth. ☐

- I know some of the effects of natural events and human activity on weather patterns. ☐

Ecology in our school grounds

1 a Clown fish are found in the coral reefs of the Pacific Ocean.

   i Suggest a reason why the clown fish are not found in British lakes.

_____ [1 mark]

   ii Suggest a reason why there are many undiscovered species in the Pacific Ocean.

_____ [1 mark]

b Wheat fields are artificial ecosystems and woods are natural ecosystems. Woods tend to have a higher biodiversity.

   i What is meant by the term **biodiversity**?

_____ [1 mark]

   ii Use your knowledge about invertebrates and vertebrates to explain how the use of pesticides could reduce the biodiversity of the wheat field.

_____

_____ [2 marks]

2 The following formula is used to estimate a population.

$$\frac{\text{number of animals caught first time} \times \text{number of animals caught second time}}{\text{number of marked animals caught second time}} = \text{population}$$

Researchers want to know the number of voles living in a wood. They set 30 traps in a small area of the wood and catch 20 voles. They mark the voles and then release them back into the wood. A week later they set more traps and catch 10 voles, five of them are marked.

a Estimate the population of voles in the wood.

_____

_____ [2 marks]

b The researchers are not convinced that their estimate is correct.

   i Suggest **two** ways in which they can improve their method.

_____

_____ [2 marks]

   ii Explain why your suggestions would make the estimate more accurate.

_____

_____ [2 marks]

Grouping organisms

1 a The picture shows a penguin.

   i Finish this sentence.

   The penguin belongs to the vertebrate group

   called _____ .　　　　[1 mark]

   ii Salmon belong to the vertebrate group called fish because they have scales. Explain why the penguin belongs to the group you have chosen.

_____

_____　　[2 marks]

b Explain why spiders are invertebrates.

_____　　[1 mark]

c This table compares animals and plants. Finish the table to describe animals.

	food	shape	movement
plants	make own	spread out	stay in one place
animals			

[3 marks]

2 A Zorse is a cross between two different species, a zebra and a horse.

  D–C

  B–A*

a What name is used to describe a cross between two different species?

_____　　[1 mark]

b Explain why the Zorse is difficult to classify.

_____　　[1 mark]

3 The lion and tiger are different species.

  D–C

a What is meant by the term **species**?

_____

_____　　[2 marks]

b Lions and tigers belong to the same family of cats. This table shows the Latin names of different cats.

common name	Latin name
bobcat	Felix rufus
cheetah	Acinonyx jubatus
lion	Panthera leo
ocelot	Felix pardalis

[3 marks]

   i Two of these cats are more closely related than the others. Write down the **common** names of these **two** cats.

_____ and _____　　[1 mark]

   ii What is the reason for your answer to part **b i**?

_____　　[1 mark]

The food factory

1 Plants make glucose ($C_6H_{12}O_6$) by a process called photosynthesis.

B–A*

a Finish the balanced symbol equation for photosynthesis.

_____ + _____ → $C_6H_{12}O_6$ + _____   [1 mark]

D–C

b The products of photosynthesis have many uses. Finish the table to describe these uses. The first one has been done for you.

product of photosynthesis	use in the plant
glucose	*energy*
cellulose	
protein	
oil	

[3 marks]

D–C

2 Light can change the rate of photosynthesis.

a Write down the names of **two other** factors that change the rate of photosynthesis.

1 _____

2 _____   [2 marks]

b The graph shows the effect of light intensity on the rate of photosynthesis.

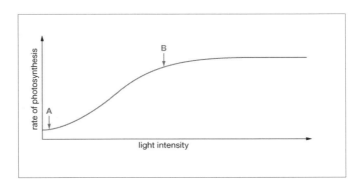

i Describe the pattern shown in the graph.

_____

_____

_____   [2 marks]

D–C

3 Glucose is a product of photosynthesis. The plant uses a process to release energy from glucose.

a Write down the name of this process.

_____   [1 mark]

b Plants carry out this process 24 hours a day. Explain why.

_____   [1 mark]

Compete or die

1 a The red and grey squirrels share the same ecological niche. What is meant by the term **ecological niche**?

_____ [1 mark]

B–A*

b The red squirrel is native to Britain. The grey squirrel was introduced to Britain about 130 years ago. During the last 100 years grey squirrel numbers have increased and red squirrel numbers have decreased.

D–C

i Suggest a reason for the change in numbers.

_____ [1 mark]

ii Grey squirrels have been removed from the island of Anglesey. Suggest the effect this might have on the red squirrel population.

_____ [1 mark]

2 This is a diagram of a food chain.

D–C

a In one year the population of shrews decreases.

i Describe the effect this would have on the badger population.

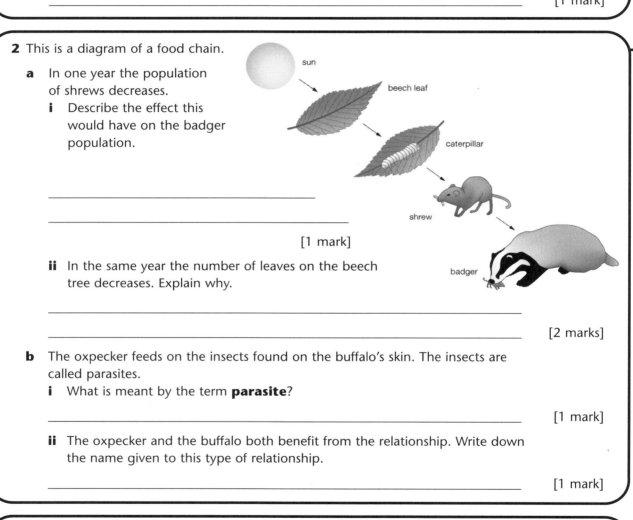

sun
beech leaf
caterpillar
shrew
badger

_____
_____

[1 mark]

ii In the same year the number of leaves on the beech tree decreases. Explain why.

_____
_____ [2 marks]

b The oxpecker feeds on the insects found on the buffalo's skin. The insects are called parasites.

i What is meant by the term **parasite**?

_____ [1 mark]

ii The oxpecker and the buffalo both benefit from the relationship. Write down the name given to this type of relationship.

_____ [1 mark]

3 Look at the graph. It shows the predator-prey relationship of lemmings and snowy owls. Write about how and why the number of lemmings change as indicated in the graph.

B–A*

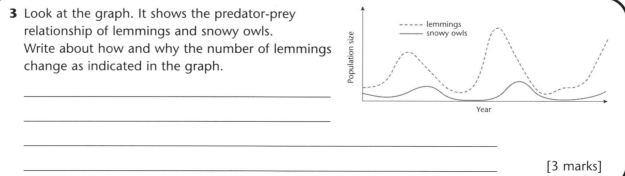

- - - - - lemmings
——— snowy owls

Population size

Year

_____
_____
_____
_____ [3 marks]

Adapt to fit

1 a Look at the picture of a camel. The table shows how the camel is adapted to help it live in the desert. Finish the table by writing how these adaptations help it to survive. The first one has been done for you.

adaptation	why it helps the camel survive
large feet	*stop sinking into sand*
no fat on body, except in hump	
hair-lined nostrils	
higher body temperatures do not harm camel	

[3 marks]

b Polar bears are adapted to live in the cold. They have thick fur for insulation as this stops them losing too much body heat.
  i Write about other ways they are adapted to live in the cold.

_____

_____

_____

_____

[3 marks]

  ii Polar bears are not in found in the same habitat as brown bears. Explain why. Use ideas about competition in your answer.

_____

_____

[2 marks]

2 a The cactus has spines instead of leaves. Explain why.

_____

_____

[2 marks]

b Lilies are pollinated by insects. Write down **one** adaptation of insect-pollinated flowers.

_____

[1 mark]

c Grasses are pollinated by wind. Descibe **two** ways they are from different insect-pollinated plants.

_____

_____

[2 marks]

Survival of the fittest

1 a Here are four sentences (**A–D**) about how the fossil of a dinosaur is formed. They are in the wrong order. Write the letters in the boxes to show the right order. The first one has been done for you.

A The dinosaur's hard parts were replaced by minerals.

B The dinosaur died.

C The dinosaur's soft tissue rotted away.

D The dinosaur became covered by sediment.

B			

[2 marks]

b The fossil record shows how living things have changed over time. The fossil record is incomplete. Explain why.

_____

_____ [2 marks]

c Creationists believe that each living thing was created individually and did not evolve. Use your knowledge about the fossil record to suggest **one** argument for and **one** argument against this theory.

For _____

_____

Against _____

_____ [2 marks]

D–C

B–A*

2 The following article gives information on the superbug MSRA. Read the article carefully and use it to help you answer the questions.

MRSA Where did it come from? MRSA evolved because of natural selection. There are lots of different strains of the bacteria. Each strain has slightly different DNA. The DNA is also constantly mutating as the bacteria reproduce. Some of these mutations will be more resistant to antibiotics than others.

When people take antibiotics, the less resistant strains die first, the more resistant strains are harder to destroy. If people stop taking the antibiotics too soon the resistant strains survive to reproduce and pass on their DNA. In this way, more and more strains evolve to be resistant to these new drugs.

a What is meant by the term **natural selection**?

_____ [1 mark]

b Use Darwin's theory of natural selection to explain how MRSA has evolved.

_____

_____ [2 marks]

D–C

B–A*

Population out of control?

1 a The rise in human population is causing an increase in the level of carbon dioxide in the air. Suggest **two** effects this increase may have on the environment.

1 _____

2 _____ [2 marks]

b The ozone layer in the Earth's atmosphere protects us from harmful ultraviolet rays. Chemicals are destroying the ozone layer.

i Write down the name of these chemicals.

_____ [1 mark]

ii The over-use of these chemicals has caused an increase in skin cancer. Suggest a reason why.

_____ [1 mark]

2 The graph shows the past, present and predicted future of the world human population.

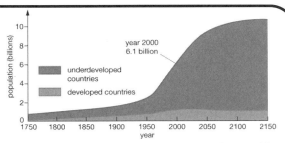

a The human population is in the rapid growth stage. What is the name given to this stage of growth?

_____ [1 mark]

b In developed countries, such as America, the population is constant. In underdeveloped countries, such as Africa, the population is still rising, yet underdeveloped countries cause less pollution. Suggest **two** reasons why.

1 _____

2 _____ [2 marks]

3 Scientists look for the water louse when they want to measure the level of pollution in water.

a Write down the name given to species that are used to measure levels of water pollution.

_____ [1 mark]

b This table shows the sensitivity of different animals to pollution.

animal	sensitivity to pollution	animal	sensitivity to pollution
stonefly larva	sensitive	freshwater mussel	semi-sensitive
water snipe fly	sensitive	damselfly larva	semi-sensitive
alderfly	sensitive	bloodworm	tolerates pollution
mayfly larva	semi-sensitive	rat-tailed maggot	tolerates pollution

A river sample contains mussels, damselfly larva and bloodworms, but no alderfly or stonefly larva. Use this information to explain why the river sample is polluted.

_____

_____ [2 marks]

Sustainability

1 a This is a picture of the dodo. The dodo no longer exists.
Suggest **two** reasons why the dodo no longer exists.

1 _____

2 _____ [2 marks]

D–C

2 Some animals are close to extinction. They are called **endangered species**.
The panda is an endangered species.

a Pandas are kept in captivity. Suggest how this may help increase the panda population.

_____ [1 mark]

b Pandas live in a remote part of China. Their natural habitat is being destroyed.
Some people want to save the panda from extinction. Suggest **two** reasons why
saving the panda will help the people who live in the same habitat.

1 _____

2 _____ [2 marks]

D–C

B–A*

3 a Some countries want to hunt whales for food. Suggest **one** argument for and
one argument against hunting whales.

For _____

_____

Against _____

_____ [2 marks]

b It is very difficult to stop people hunting whales.
Suggest **one** reason why.

_____ [1 mark]

D–C

B–A*

4 a Sustainable resources are resources that should not run out. Fish are a sustainable
resource. The government has set fish quotas. Fishermen can only catch so many
fish at any one time. The fish quotas should help maintain the population of fish in
the sea. Explain why.

_____

_____ [2 marks]

b As the human population increases there will be more demand for energy
resources. The sustainable development of fast-growing willow trees could help
solve the problem. Explain why.

_____

_____ [2 marks]

D–C

B–A*

B2 Revision checklist

- I know how to collect and use data to estimate a population. ☐

- I know how to use a key to identify plants and animals. ☐

- I know the characteristics of the different vertebrate groups. ☐

- I can state the word and formula equation for photosynthesis. ☐

- I can describe the effect of increased light, temperature and carbon dioxide on photosynthesis rate. ☐

- I can explain how similar organisms compete for the same ecological niche. ☐

- I can explain how the size of a predator population will affect the prey population. ☐

- I can explain how adaptations of organisms determine their distribution and abundance. ☐

- I can explain how camels, polar bears and cacti are adapted to their habitats. ☐

- I can describe how organisms became fossilised. ☐

- I know that developed countries have a greater effect on world pollution. ☐

- I can explain the effects of increased pollution on global warming, acid rain and the ozone layer. ☐

- I can describe ways in which animals become extinct. ☐

- I can explain sustainable development and describe how it may protect endangered species. ☐

Paints and pigments

1 a An **emulsion paint** is a water-based paint. Explain how it covers a surface.

_____

_____ [2 marks]

D–C

b Sam and Chris are discussing why paint is a colloid. Sam gives an explanation. He uses ideas about particle size and mixtures and dispersion. Write down what he says.

_____

_____

_____

_____ [4 marks]

B–A*

c The oil in oil paint is very sticky and takes a long time to harden. Explain how it hardens in two stages and what it forms.

_____

_____

_____ [3 marks]

2 a Thermochromic pigment changes colour at 45 °C. Write down **two** examples it is used for.

_____

_____ [2 marks]

D–C

b Most thermochromic pigments change from a particular colour to colourless. How do they ensure a greater range of colours is available?

_____ [1 mark]

B–A*

c Thermochromic paints come in a limited range of colours. If a green paint becomes yellow when heated explain what the mixture contains and how it changes colour.

_____

_____

_____

_____ [4 marks]

d A pigment that stores absorbed energy is called

_____ [1 mark]

D–C

e What does it release that energy as?

_____ [1 mark]

3 a Where are phosphorescent pigments used?

_____ [1 mark]

B–A*

b What have phosphorescent pigments replaced and why?

_____

_____ [2 marks]

Construction materials

D–C

1 a Put these materials into the order of hardness.

granite   **limestone**   **marble**

Least hard _____ _____ _____ Hardest   [3 marks]

b Brick, concrete, steel, aluminium and glass are manufactured.
Finish the table to show the raw materials they come from.

building material	brick	cement	glass	iron	aluminium
raw material					

[5 marks]

B–A*

c **Igneous**, **sedimentary** and **metamorphic** rocks differ in hardness.

Which rock is usually the least hard? _____ [1 mark]

d Granite and marble are different types of rock. Their hardness is different. Compare how granite and marble are made and why their structures are different. Use **all** the words in this list.

crystals   **igneous**   **interlocking**   **heat**   **pressure**   **slowly**   **metamorphic**   **solidifies**

granite	marble

[8 marks]

e Limestone is a **sedimentary rock**. Explain how limestone is made.

_____

_____ [3 marks]

D–C

2 a Calcium carbonate decomposes at a very high temperature. Write a word equation.

_____ [3 marks]

b Cement is made from limestone. Write down how.

_____ [2 marks]

B–A*

c Write down the symbol equation for the thermal decomposition of calcium carbonate.

_____ [3 marks]

d Reinforced concrete is a better construction material than non-reinforced concrete. Explain why. Use ideas about stretching and tension in concrete and strength under tension of steel.

_____

_____

_____

_____ [4 marks]

Does the Earth move?

1 a Are the tectonic plates that make up the Earth's crust less dense or more dense than the mantle?

_____ [1 mark]

D–C

b Write down the **two** kinds of tectonic plate.

_____ and _____ [2 marks]

c The mantle is always solid, but at greater depths it is more like Plasticine, which can 'flow'. The **tectonic plates** move very slowly on this mantle. They can move in different ways. Explain how, using diagrams to help you explain.

B–A*

[5 marks]

d Explain how subduction occurs.

_____
_____
_____
_____
_____ [6 marks]

e Write down **two** pieces of evidence used to develop the theory of plate tectonics. Use ideas about continental drift and mid-ocean ridges.

_____
_____
_____ [4 marks]

2 a Magma can rise through the Earth's crust. Explain why. [1 mark]

D–C

_____

b Magma cools and solidifies into igneous rock either after it comes out of a volcano as lava, or before it even gets to the surface. By looking at crystals of igneous rock, geologists can tell how quickly the rock cooled. Fill in the **two** boxes with an explanation and an example.

basalt   **cools rapidly**   **cools slowly**   **granite**

small crystals

large crystals

[4 marks]

c i There are two types of magma, iron-rich and silica-rich. Which is the magma that is less runny and produces volcanoes that may erupt explosively?

B–A*

_____ [1 mark]

ii Explain what it produces.

_____
_____ [2 marks]

Metals and alloys

Grades

D–C

1 a Copper used for recycling has to be sorted carefully so that valuable 'pure' copper scrap is not mixed with less pure scrap. When impure copper is used to make alloys what must happen first?

_____ [1 mark]

b If the scrap copper is very impure what must be done before it is used again?

_____ [1 mark]

D–C

2 Impure copper can be purified in the laboratory using an electrolysis cell.

a What is the **anode** made from?

_____ [1 mark]

b What happens at the **cathode**?

_____ [1 mark]

B–A*

c What happens to the **anode**?

_____ [1 mark]

d What happens to the thickness of the cathode?

_____ [1 mark]

e What is the impure copper called?

_____ [1 mark]

f What happens to the impurities from the copper anode?

_____ [1 mark]

D–C

3 a Most metals form alloys. Draw a **straight** line to match the metals to the alloy.

amalgam	contains copper and zinc
solder	contains mercury
brass	contains lead and tin

[2 marks]

b Alloys are often more useful than the original metals, though nowadays pure copper is more important than bronze or brass. Why are vast amounts turned into electric wire?

_____ [1 mark]

B–A*

c Write down **two** properties of smart alloys.

_____
_____ [2 marks]

d Write down **two** new ways of using smart alloys.

_____
_____ [2 marks]

e **Nitinol** is a smart alloy. Which **two** metals is it made from?

_____
_____ [2 marks]

43

Cars for scrap

1 a In winter, icy roads are treated with salt. Why is this a problem for steel car bodies?

_____ [1 mark]

b Aluminium does not corrode in moist air. Explain why.

_____ [1 mark]

c Rust is an oxide layer but it does not protect the rest of the iron. Explain why.

_____ [1 mark]

d Rusting is a **chemical reaction** between iron and oxygen to make an oxide.
The chemical name for rust is **hydrated iron(III) oxide**.
Write down the word equation for rusting.

_____ [3 marks]

D–C

B–A*

---

2 a Steel is an alloy made of _____ and _____. [2 marks]

b Write down **two** advantages of steel over iron.

_____ [2 marks]

c Steel and aluminium can both be used to make car bodies but each material has its own advantages. Write down **two** advantages of cars made from aluminium and **one** disadvantage.

_____

_____

_____ [3 marks]

d Finish this table. Write down the reasons why these materials are used.

material and its use	reasons material is used
aluminium in car bodies and wheel hubs	
copper in electrical wires	
plastic in dashboards, dials, bumpers	
pvc in metal wire coverings	
plastic/glass composite in windscreens	
fibre in seats	

[6 marks]

D–C

B–A*

---

3 a Write down **three** benefits that recycling metals and the other materials of a car have on the environment.

_____

_____

_____ [3 marks]

D–C

Clean air

D–C

1 a i Label the pie chart with the four main gases of the air in the correct section.

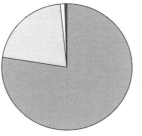

[4 marks]

ii Mark in the percentages of the gases.

[4 marks]

iii These percentages do not change very much because there is a balance between three of the processes that use up or make carbon dioxide and use or make oxygen. Explain how the balance is maintained.

_____

_____

_____

_____ [7 marks]

B–A*

2 a Over the last few centuries the percentage of carbon dioxide in air has increased slightly. This is due to a number of factors. Explain these factors.

increased energy usage	
increased population	
deforestation	

[3 marks]

b Explain **one** theory of how the Earth's atmosphere has evolved. Include in your answer
  i the original gases of the atmosphere
  ii how nitrogen levels increased
  iii how the levels of carbon dioxide and oxygen evolved to those which they are today.

_____

_____

_____ [4 marks]

D–C

3 a Finish the table to describe the origin of these atmospheric pollutants.

pollutant	carbon monoxide	oxides of nitrogen	sulphur dioxide
origin of pollutant			

[3 marks]

B–A*

b Which metal catalyst do catalytic converters contain?

_____ [1 mark]

c i A reaction between nitric oxide and carbon monoxide takes place on the surface of the catalyst. The reaction forms nitrogen and carbon dioxide. Why is it important that these two gases are made?

_____ [1 mark]

ii Write down the word equation for this reaction.

_____ [2 marks]

iii Write down the balanced symbol equation for this reaction.

_____ [3 marks]

Faster or slower (1)

1 a Look at **Graph A**. It shows the reaction between magnesium and acid at 20 °C.

i If the reaction takes place at a higher temperature mark the reaction line that you would expect on the graph.

[2 marks]

ii Which graph has the steeper gradient?

_____

_____

[1 mark]

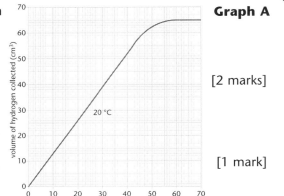
Graph A

iii The reaction rate increases at higher temperatures. Explain why. Use ideas about particles in your answer.

_____

_____

[4 marks]

b The rate of a reaction is not determined by the number of collisions. Explain what it is determined by.

_____

_____

[2 marks]

c Explain the increase in reaction rate in terms of particle collisions.

_____

[1 mark]

2 Look at **Graph B**. It shows the reaction between magnesium and acid of different concentrations.

a What does the graph show about

i the reaction rates with the two acids?

_____

_____

ii the amount of hydrogen collected in the two reactions?

_____

_____

[4 marks]

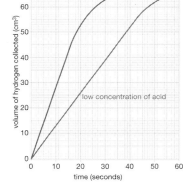
Graph B

b The rate of reaction can be worked out from the gradient of **Graph C**.

i Draw construction lines on the graph and calculate the gradient of the line.

_____

[3 marks]

ii What does **interpolation** mean?

_____

_____

[1 mark]

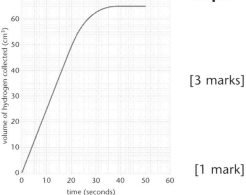
Graph C

Faster or slower (2)

D–C

1 a What products are made during an explosion?

_____ [2 marks]

D–C

2 a The reaction between calcium carbonate and hydrochloric acid is measured by the decrease in mass. Look at the equation, why is there a decrease in mass?

$$CaCO_3 + 2HCl \rightarrow CaCl_2 + H_2O + CO_2$$

_____ [2 marks]

b The graph shows how the rate of reaction between calcium carbonate and dilute hydrochloric acid is measured.

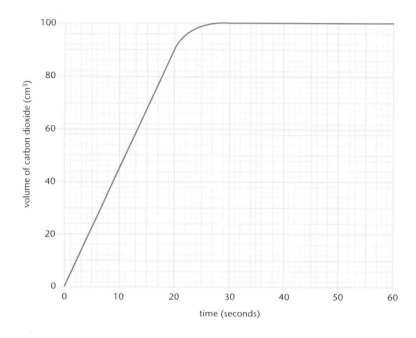

i At which time does the reaction stop? _____ [1 mark]

ii If this reaction were using powdered calcium carbonate, sketch on the graph the line that would show the reaction of the same mass of calcium carbonate as lumps. [2 marks]

iii Why does the reaction slow down? Use ideas about collisions in your answer.

_____ [1 mark]

B–A*

c Why does the reaction rate increase if the surface area increases. Use ideas about collisions between particles in your answer.

_____ [1 mark]

D–C

2 a What does a catalyst do?

_____ [1 mark]

B–A*

b Most catalysts only make a **specific** reaction faster. Explain why.

_____

_____ [2 marks]

C2 Revision checklist

- I know that paint is a colloid where solid particles are dispersed in a liquid, but are not dissolved. ☐

- I know that thermochromic pigments change colour when heated or cooled. ☐

- I know that brick is made from clay, glass from sand, and aluminium and iron from ores. ☐

- I know that the decomposition of limestone is:
  calcium carbonate → calcium oxide + carbon dioxide. ☐

- I know that the outer layer of the Earth is continental plates with oceanic plates under oceans. ☐

- I know that the size of crystals in an igneous rock is related to the rate of cooling of molten rock. ☐

- I know that copper can be extracted by heating its ore with carbon, but purified by electrolysis. ☐

- I know that alloys are mixtures of metals, for example, copper and zinc make brass. ☐

- I know that aluminium does not corrode when wet as it has a protective layer of aluminium oxide. ☐

- I know that iron rusts in air and water to make hydrated iron(III) oxide. ☐

- I know that toxic carbon monoxide comes from incomplete combustion of petrol or diesel in cars. ☐

- I know that catalytic converters can remove CO by conversion:
  $2CO + 2NO \rightarrow N_2 + CO_2$. ☐

- I know that a temperature increase makes particles move faster, so increasing the rate of reaction. ☐

- I know that a catalyst is a substance which changes the rate of reaction but is unchanged at the end. ☐

Collecting energy from the Sun

D–C

1 a Write down **four** advantages of using photocells.

_____

_____

_____

_____ [4 marks]

B–A*

b A p-n junction is made from two pieces of silicon. Explain how the two pieces of silicon are different and what causes the difference.

_____

_____

_____ [3 marks]

D–C

2 a During the day, energy from the Sun passes through the large window and warms the room; this is called passive solar heating.
How is the room heated during the night?

Day

_____ [1 mark]

B–A*

b The diagram represents the electromagnetic spectrum.

X-rays	ultraviolet	visible light	infrared	radio

i Write the letter **S** on the diagram to show the wavelength of radiation from the Sun that is absorbed by plants in a greenhouse.

ii Write the letter **P** on the diagram to show the wavelength of radiation that is re-radiated from the plants in a greenhouse. [2 marks]

D–C

3 a A wind turbine transfers the kinetic energy of the wind into electricity.
What is wind?

_____ [1 mark]

B–A*

b Write down **two** advantages and **two** disadvantages of generating electricity using wind turbines.

Advantages _____

_____

Disadvantages _____

_____ [4 marks]

Generating electricity

1 a The diagram shows a model dynamo. When the coil is spun, a current is produced. Write down **two** ways in which the size of the current can be increased.

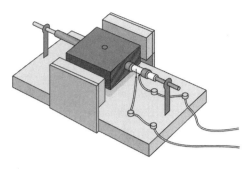

_____

_____ [2 marks]

b A model generator consists of a coil of wire rotating between the poles of a magnet. How is the structure of a generator at a power station different from the model generator?

_____ [1 mark]

2 The flow diagram represents how fossil fuels at a power station provide electrical energy for distribution around the country.

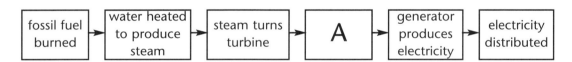

fossil fuel burned → water heated to produce steam → steam turns turbine → **A** → generator produces electricity → electricity distributed

a One step in the process has been missed out. What should be in box **A**?

_____ [1 mark]

b High voltage transmission lines distribute electricity around the country at 400 kV. We use electricity in our homes at 230 V. Explain why electricity is **not** distributed at 230 V.

_____

_____

_____ [3 marks]

c A power station is about 30% efficient. What are the main causes of inefficient energy transfer in a power station?

_____

_____

_____ [3 marks]

Fuels for power

D–C

1 a A nuclear power station uses uranium as its energy source. Uranium is not a fuel.

   i What is a fuel?

   _____ [2 marks]

   ii How does uranium provide energy in the form of heat?

   _____ [1 mark]

B–A*

b Biomass is an energy source that produces carbon dioxide as a waste product.

   i Suggest a problem that is caused by carbon dioxide.

   _____ [1 mark]

   ii Write down the advantage of using biomass as an energy source.

   _____ [1 mark]

D–C

2 a Each of the headlamp bulbs in Anna's car is connected to a 12 V battery. When she switches on the headlamps, a current of 2 A passes through the bulb. Calculate the power rating of the bulb.

   _____ [4 marks]

b In her home, Anna uses a 2.5 kW kettle for $\frac{1}{2}$ hour each day. Electricity costs 12p per kWh. How much does it cost Anna to use her kettle each day?

   _____ [4 marks]

B–A*

c Anna sets her dishwasher to work overnight. Why is electricity cheaper during the night?

   _____ [1 mark]

D–C

3 a Finish the sentences by choosing the **best** words from this list.

   calcium   cancer   DNA   kinetic   nuclear   mutate   react

   Radiation from _____ energy sources causes ionisation.

   This causes a change in the structure of atoms. One of the chemicals in body

   cells is _____ and when this is exposed to radiation it can

   _____ . As a result, body cells may divide in an uncontrolled

   way. This can lead to _____ . [4 marks]

B–A*

b Write down the advantages and disadvantages of using uranium as an energy source.

   _____

   _____

   _____ [4 marks]

Nuclear radiations

1 a Background radiation is always present. Write down **two naturally occurring** sources of background radiation.

_____

_____ [2 marks]

b Answer **true** or **false** to each of the following statements about alpha, beta and gamma radiation.

Gamma radiation causes more ionisation than alpha radiation.	
Alpha radiation has a range of a few centimetres in air.	
Beta radiation comes from the nucleus of an atom.	
Beta radiation can be absorbed by a thin sheet of paper.	

[4 marks]

c Atoms are neutral because they contain the same number of positive protons and negative electrons. Explain how negative and positive ions are formed when an atom is exposed to radiation.

Negative ions _____

_____

Positive ions _____

_____ [2 marks]

2 Gamma radiation is used to sterilise medical instruments. It also has other medical uses.

a Write down **one other** medical use for gamma radiation.

_____ [1 mark]

b What property of gamma radiation makes it suitable?

_____ [1 mark]

3 Radioactive waste must be stored securely for possibly thousands of years.

a Why must it be stored for so long?

_____ [1 mark]

b Some people are worried that terrorists may make a nuclear bomb from nuclear waste. Discuss whether or not we should have any concerns about terrorists obtaining nuclear waste.

_____

_____

_____

_____

_____ [3 marks]

Our magnetic field

D–C

1 a Which statement best describes the magnetic field due to a coil of wire?
Put a tick (✓) in the box next to the correct answer.

The magnetic field is circular.	
The magnetic field is radial like the spokes of a bicycle wheel.	
The magnetic field is similar to the field due to a bar magnet.	

[1 mark]

B–A*

b Earth's outer core is mainly molten iron but it is too hot to be magnetic.
Suggest how Earth's magnetic field is produced.

_____

_____

_____ [3 marks]

D–C

2 a Scientists believe that the Moon and Earth used to rotate much faster than they
do now. What do they think caused the rotation to slow down?

_____ [1 mark]

B–A*

b The Moon and Earth were probably formed at the same time. Explain how
scientists believe the Moon and Earth were formed.

_____

_____

_____

_____

_____ [3 marks]

B–A*

3 Cosmic rays originate from the Sun. What happens when they interact with Earth's
magnetic field?

_____

_____ [2 marks]

4 Finish the sentences by choosing the **best** words from this list.

charged particles   **gamma rays**   **hydrogen atoms**   **radio signals**

D–C

Solar flares emit _____ that produce

magnetic fields. The magnetic fields interact with Earth's magnetic field.

B–A*

This causes interference with _____ .

[2 marks]

Exploring our Solar System

1 a On August 24 2006, the International Astronomical Union considered a proposal to redefine planets. There would be twelve planets in our Solar System. Ceres, the largest asteroid would become a planet. Charon, at the moment known as Pluto's moon, would become a 'twin planet' with Pluto. Recently another planet has been discovered beyond Pluto. This proposal was rejected and the decision made that Pluto should no longer be called a planet.

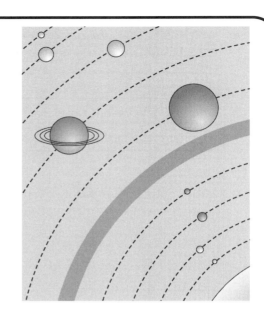

The diagram shows the Sun, the eight planets, Pluto and the asteroid belt.

i Write the letter **C** on the diagram to show where Ceres orbits.

ii Write the letter **P** on the diagram to show Pluto.

[2 marks]

b The diagram represents the Moon in orbit around Earth.
  i What is the name of the force that keeps the Moon in orbit around Earth?

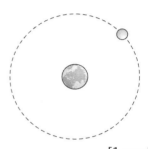

_____

_____

[1 mark]

ii Add an arrow to the diagram to show the direction in which this force acts on the Moon.

_____

[1 mark]

D–C

B–A*

---

2 a Helen Sharman was the first British astronaut. She spent eight days in the Mir Space Station doing science experiments. When she was in the Space Station she experienced **weightlessness** but was not **weightless**.
Why was Helen never weightless?

_____ [1 mark]

b When astronauts work outside a spacecraft, they have to wear special helmets with sun visors. Why do the helmets need special visors?

_____ [1 mark]

c NASA is planning to send a manned spacecraft to another planet before 2030. It is expected to cost £400 billion.
Which planet will the spacecraft go to? Explain the reason for your answer.

_____

_____ [2 marks]

D–C

B–A*

---

3 Proxima Centauri is 4.22 **light-years** away from us.
What is meant by the term **light-year**?

_____ [1 mark]

B–A*

Threats to Earth

1 a Most asteroids orbit the Sun in a belt between two planets. Which two planets?

_____ and _____

[2 marks]

b What are asteroids? Put a tick (✓) in the box next to the correct answer.

clouds of very dense gas	
cubes of ice	
rocks that have broken away from planets	
rocks left over when the Solar System formed	

[1 mark]

c Asteroids have not joined together to form another planet. Explain why.

_____

_____ [2 marks]

2 The diagram shows the orbits of two bodies orbiting the Sun.
One is a planet, the other is a comet.

a Label the orbit of the comet with the letter **C**.

[1 mark]

b Write the letter **X** to show where on the orbit the comet is travelling at its fastest. [1 mark]

c Why does a comet's tail always point away from the Sun?

_____ [1 mark]

3 Scientists are constantly updating information on the paths of NEOs.

a Why is it important to constantly monitor the paths of **NEO**s?

_____

_____ [2 marks]

b Scientists may want to change the course of a NEO. Explain how this may
be done.

_____

_____

_____ [2 marks]

The Big Bang

1 a The Universe is expanding. Galaxies in the Universe are moving at different speeds. Which galaxies are moving the fastest?

_____ [1 mark]

b Scientists know how fast galaxies are moving because they measure **red shift**. What is meant by 'red shift'?

_____
_____
_____ [3 marks]

c How does red shift provide information about the speed of a galaxy?

_____
_____ [1 mark]

2 a A star starts its life as a swirling cloud of gas and dust. Describe what happens to this cloud to produce a glowing star.

_____
_____
_____
_____
_____ [4 marks]

b A star's life is determined by its size.
  i What happens to a medium-sized star, like our Sun, at the end of its life?

_____
_____
_____
_____
_____ [4 marks]

  ii What happens to a very large star at the end of its life?

_____
_____
_____
_____
_____ [4 marks]

P2 Revision checklist

- I can explain how energy from the Sun can be used for heating and producing electricity. ☐

- I can describe how a p-n junction produces electricity. ☐

- I can describe how generators produce electricity. ☐

- I can explain why electricity is distributed via the National Grid at high voltages. ☐

- I can explain the advantages and disadvantages of the fuels used in power stations. ☐

- I can calculate power and the cost of using an electrical appliance for a certain time. ☐

- I know why there is background radiation. ☐

- I can list some uses of alpha, beta and gamma sources and relate their use to their properties. ☐

- I can explain how the Earth's magnetic field is produced and its effect on cosmic rays. ☐

- I can describe how the Moon was formed. ☐

- I can explain why the planets stay in the orbits they do around the Sun. ☐

- I can explain how and why we are exploring space through manned and unmanned spacecraft. ☐

- I can describe asteroids and comets and know the importance of constantly checking NEOs. ☐

- I can describe the Big Bang theory and why scientists believe the Universe is still expanding. ☐

Molecules of life

1 Respiration is a chemical reaction that takes place in the cell.
Write down the name of the part of the cell where respiration takes place.

_____ [1 mark]

D–C

2 This diagram shows DNA fingerprints of individuals
connected to a robbery.

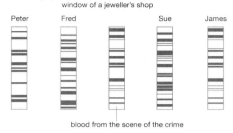

DNA fingerprints from suspects' blood left at the broken window of a jeweller's shop

Peter   Fred   Sue   James

blood from the scene of the crime

a Who left blood at the crime scene? Explain your answer.

_____

_____

_____

_____

[2 marks]

b Describe how a DNA fingerprint is made.

_____

_____

_____

[3 marks]

c The DNA base code codes for amino acids.
  i How many amino acids are coded for in the following section of DNA?
  TATATGTAAAAACAA

_____ [1 mark]

  i Write down the complementary base sequence for this section of DNA.

_____ [1 mark]

D–C

B–A*

3 Look at the graph. It shows the effect of
temperature on an enzyme.

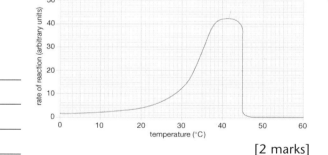

a What is meant by the term **enzyme**?

_____

_____

_____

_____

[2 marks]

b Describe the pattern shown in the graph.

_____

_____

[2 marks]

c What is the optimum temperature of this enzyme?

_____

[1 mark]

d The enzyme controlled reaction stops at 45 °C. Use ideas about enzyme shape
to explain this.

_____

_____

_____

[3 marks]

D–C

B–A*

Diffusion

Grades

D–C

1 a What is meant by the term **diffusion**?

_____

_____
[2 marks]

B–A*

b Diffusion takes place between cells of the body.
Describe **three** ways to increase the rate of diffusion between cells.

1 _____

2 _____

3 _____ [3 marks]

D–C

2 a Diffusion takes place in the placenta. Substances diffuse from the foetus into the
mother's blood. Write down the name of **two** of these substances.

1 _____

2 _____ [2 marks]

b Diffusion of oxygen into the blood takes place in the lungs.
In which part of the lung does oxygen enter the blood?

_____ [1 mark]

D–C

3 Describe how carbon dioxide moves into the leaf. Use the words **concentration**,
diffusion and **photosynthesis** in your answer.

_____

_____

_____ [3 marks]

B–A*

4 Look at the picture of a synapse. Describe how the
synapse is adapted to carry the signal from one neurone
to the next.

_____

_____

_____

_____

_____

_____
[3 marks]

Keep it moving

1 Red blood cells are adapted to carry out their function. They are disc-shaped and do not have a nucleus.

D–C

    a Explain how these adaptations allow them to support their function.

       Disc-shaped _____

       _____

       No nucleus _____

       _____ [2 marks]

    b Write down the name of the chemical that makes red blood cells red.

       _____ [1 mark]

2 a Look at the diagram of the heart.

D–C

      i Draw an **X** to show the part of the heart that receives blood from the lungs.

      ii Label the bicuspid valve.   RIGHT   LEFT   [2 marks]

    b The left ventricle has a thicker wall than the right ventricle. Explain why.

       _____

       _____ [2 marks]

    c People with heart disease may need a heart transplant.

B–A*

      i Describe **two** problems with heart transplants.

       1 _____

       2 _____ [2 marks]

      ii Describe **one** advantage and **one** disadvantage that heart transplants have over heart pacemakers.

       Advantage _____

       Disadvantage _____ [2 marks]

3 a Describe the role of blood vessels in circulating blood around the body.

D–C

       _____

       _____

       _____ [3 marks]

    b Humans have a double circulatory system. Explain the advantage of a double circulatory system.

B–A*

       _____

       _____ [2 marks]

Divide and rule

1 a Amoebas are unicellular organisms. They only have one cell. Humans are multi-cellular; they are made of many cells. It may be a disadvantage to be unicellular rather than multi-cellular. Explain why.

_____

_____ [2 marks]

b The table shows the surface area and volume of different cubes.
   i Finish the table by calculating the surface area to volume ratio of each cube. The first one has been done for you.

cube	surface area in cm^2	volume in cm^3	ratio
A	24	8	24/8 = 3
B	54	27	
C	96	64	
D	150	125	

[3 marks]

   ii Organisms made of a single large cell have a disadvantage. Use the information in the table to explain why.

_____

_____ [2 marks]

2 a Write down the name of the type of cell division that makes new **body** cells.

_____ [1 mark]

b Which **two** of the following statements relate to the cell division that makes human body cells? Put ticks (✓) in the **two** correct boxes.

the new cells are diploid ☐
four new cells are made ☐
the new cells contain 23 chromosomes ☐
pairs of chromosomes separate to opposite poles of the cell ☐
the new cell shows variation ☐
chromosomes separate to opposite poles of the cell ☐   [2 marks]

3 a Sperm cells have a structure called an acrosome. Explain why sperm cells need an acrosome.

_____

_____ [2 marks]

b A special type of cell division makes sperm cells.
   i Write down the name of this type of cell division.

_____ [1 mark]

   ii Describe one way in which this type of cell division is different from the cell division that makes body cells.

_____ [1 mark]

c Sperm cells are haploid. Explain is meant by the term **haploid**.

_____ [1 mark]

Grades
D–C
B–A*
D–C
B–A*
D–C
B–A*
D–C

Growing up

1 a Both animal and plant cells have a nucleus which makes them similar.
Describe **one other** way that they are similar and **one** way they are different.

Similar _____

_____

Different _____

_____ [2 marks]

b For a fertilised egg to grow into an embryo the cells need to divide
and change.

i Write down the name that is given to cells before they become specialised.

_____ [1 mark]

ii Damaged brain cells cause a disease called Parkinson's. Scientists hope to
repair the damage by taking non-specialised cells from human embryos and
then turning them into brain cells.
Explain why some people may object to this process.

_____

_____ [2 marks]

2 a Look at the table. It shows the change in weight of a baby from 0 to 30 months.

age in months	0	3	6	9	12	15	18	21	24	27	30
weight in kg	2.5	5.0	6.4	7.5	8.8	9.6	9.8	10.0	10.1	10.4	10.7

i Plot the points on the graph. The first two have been done for you.

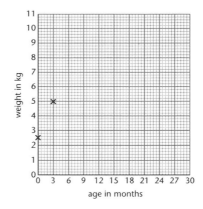

[3 marks]

ii Finish the graph by drawing the best curve. [1 mark]

iii Describe the pattern in the graph.

_____

_____ [2 marks]

iv Write down the phase of human growth shown in the graph.

_____ [1 mark]

b A baby should weigh about 14 kg at 30 months.
Suggest a reason why the baby represented in the graph is underweight.

_____ [1 mark]

Controlling plant growth

D–C

1 a Ben grows apple trees. He decides to take some shoot cuttings from his apple trees and uses rooting powder to grow new trees.
What effect does rooting powder have on the shoot cuttings?

_____ [1 mark]

b Ben also grows wheat and sprays the wheat with selective weedkiller.
The weedkiller destroys the weeds with broad leaves but not the crops.
  i How does the weedkiller destroy the weeds?

_____
_____ [2 marks]

  ii Explain why the crops are not affected by the weedkiller.

_____
_____ [2 marks]

2 a

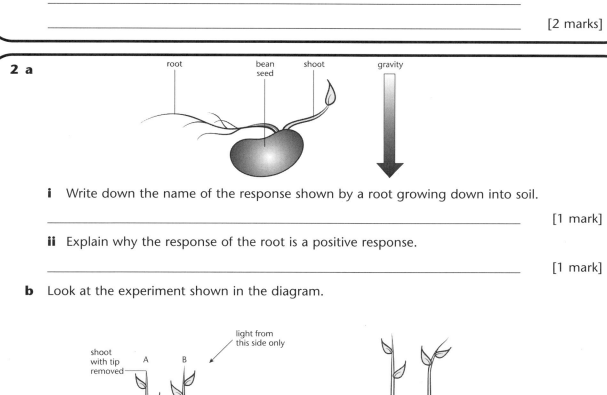

root   bean   shoot   gravity
              seed

  i Write down the name of the response shown by a root growing down into soil.

_____ [1 mark]

D–C

  ii Explain why the response of the root is a positive response.

_____ [1 mark]

B–A*

b Look at the experiment shown in the diagram.

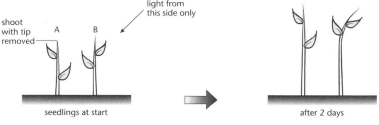

light from
this side only

shoot
with tip   A   B
removed

seedlings at start   after 2 days

  i Write down the name of the hormone that caused the growth of the **shoot**.

_____ [1 mark]

  ii How does the hormone cause **shoot B** to bend?

_____
_____
_____ [3 marks]

  iii Explain why **shoot A** did not bend.

_____
_____ [2 marks]

New genes for old

1 Richard uses selective breeding to produce apples that are resistant to disease and green in colour. Describe the process of selective breeding.

_____

_____

_____ [3 marks]

D–C

2 When a gene mutates the base sequence changes. Look at the original base code and its mutation.

original base code   **AAAGGTCACTTGAAA**

mutation   **AAAGGTCACTGTAAA**

a Describe the mutation in this base code.

_____ [1 mark]

b Look at the flow chart. It shows how a white pigment is turned into a purple pigment using two enzymes.

enzyme A   enzyme B

white pigment ⟶ red pigment ⟶ purple pigment

The reactions take place in the petals of a flower. A mutation to the DNA resulted in red flowers. Use ideas about **base sequence**, **proteins** and **enzymes** to explain this.

_____

_____

_____ [3 marks]

B–A*

3 Beta-carotene is needed to produce vitamin A. Beta-carotene is found in carrots but not in rice.

a Describe how genetic engineering can be used to grow rice that provides beta-carotene.

_____

_____

_____ [3 marks]

b Suggest **one** advantage and **one** disadvantage of genetic engineering.

Advantage _____

_____

Disadvantage _____

_____ [2 marks]

D–C

More of the same

D–C

1 a Describe how cows are cloned using **embryo transplants**.

_____

_____

_____ [3 marks]

b Scientists are hoping to solve organ transplant problems by cloning pigs.
  i Explain how the cloning of pigs could help solve organ transplant problems.

_____

_____ [2 marks]

  ii Suggest **one** reason why people may object to this process.

_____ [1 mark]

B–A*

c Look at the diagram. It shows how Dolly the sheep was cloned.
  i Describe **two** ways in which nuclear transfer is different from embryo transplants.

egg cell taken from sheep A and nucleus removed

cells taken from the udder of sheep B and the nucleus removed

nucleus from sheep B is put into egg of sheep A

egg cell is put into a female sheep to grow

cell grows into a clone of sheep B

  1 _____

  _____

  _____

  2 _____

  _____

  _____ [2 marks]

  ii The lamb produced is a clone of which sheep, A or B? Explain your answer.

_____ [1 mark]

D–C

2 Tissue culture can be used to clone plants.
  a Suggest **one** advantage and **one** disadvantage of cloning plants.

  Advantage _____

  _____

  Disadvantage _____

  _____ [2 marks]

B–A*

  b Cloning plants is easier than cloning animals. Explain why.

_____

_____

_____ [3 marks]

B3 Revision checklist

- I can interpret data on DNA fingerprinting for identification. ☐

- I can describe DNA replication. ☐

- I know that food and oxygen diffuse across the placenta. ☐

- I can describe diffusion as the net movement of particles from a region of high to low concentration. ☐

- I know that arteries transport blood away from the heart. ☐

- I know that a patient can reject a heart transplant. ☐

- I know that at fertilisation haploid gametes join to form a diploid zygote. ☐

- I know that body cells are made by mitosis and gametes are made by meiosis. ☐

- I can identify the main stages of human growth. ☐

- I know that shoots are positively phototropic and negatively geotropic; roots are the opposite. ☐

- I can describe the stages in selective breeding. ☐

- I know that genetic engineering is used to make insulin. ☐

- I can describe some advantages and disadvantages of cloned plants. ☐

- I know that cloned animals could be used to produce organs for transplants. ☐

What are atoms like?

D–C

1 a What are the particles in the nucleus of an atom?

_____ [1 mark]

b Finish the table to show the relative mass and charge of the atomic particles.

	relative charge	relative mass
electron		0.0005 (zero)
proton	+1	
neutron		

[4 marks]

c What is the atomic number of an atom?

_____ [1 mark]

d What is the mass number?

_____ [1 mark]

B–A*

e What is the atom that has an atomic number of 9, and a mass number of 19, and a neutral charge?

atomic number	9
mass number	19
charge	0
name	

f If an element has the symbol $^{35}_{17}Cl$ how many protons are there in the atom? How many neutrons?

i Protons _____ [1 mark]

ii Neutrons _____ [1 mark]

D–C

2 a What is an isotope?

_____ [1 mark]

B–A*

b Isotopes of an element have different numbers of neutrons in their atoms. Finish the table.

isotope	electrons	protons	neutrons
$^{12}_{6}C$	6	6	
$^{14}_{6}C$	6	6	

[2 marks]

B–A*

3 Draw the electronic structure for the element aluminium. Explain why **three** shells are needed for electrons.

_____

_____

_____ [5 marks]

Ionic bonding

1 a Put a tick (✓) in the box next to the sentence that describes a **metal** atom.

D–C

    i An **atom** that has extra electrons in its outer shell and needs to **lose** them to be stable. ☐

    ii An **atom** that has 'spaces' in its outer shell and needs to **gain** electrons to be stable. ☐

    [1 mark]

b Draw a diagram to show how the electrons transfer between a metal atom and a non-metal atom to form a stable pair. **Outer shells only.**

    ○   ○

    [3 marks]

c Finish the sentences.

    i If an atom loses electrons a _____ **ion** is formed.   [1 mark]

    ii An example of an atom which loses 2 electrons is _____.   [1 mark]

d Finish the sentences.

    i A **negative ion** is formed by an atom _____ electrons.   [1 mark]

    ii An example of an atom gaining 1 electron is _____.   [1 mark]

e Finish the sentences.

During **ionic bonding**, the metal atom becomes a _____

ion and the non-metal atom becomes a _____ ion.

The positive ion and the negative ion then attract one another. They attract to

a number of other ions to make a solid _____.   [3 marks]

f Draw the '**dot and cross' model** for sodium chloride. **Outer shell electrons only.**

B–A*

    [4 marks]

g Draw the dot and cross diagram for magnesium chloride.

    [4 marks]

h Put a tick (✓) in the boxes next to the substances that conduct electricity.

D–C

    sodium chloride solution ☐   solid sodium chloride ☐

    molten (melted) magnesium oxide ☐   solid magnesium oxide ☐

    molten sodium chloride. ☐   [3 marks]

i Finish the sentences for some of the physical properties of sodium chloride and magnesium oxide.

B–A*

    i There is a strong attraction between positive and negative ions so they have _____ melting points.

    ii Its ions cannot move in the solid so it does not _____.

    iii Solutions or molten liquids conduct electricity because _____.   [3 marks]

Covalent bonding

1 a Non-metals combine together by **sharing** electrons. What is this type of bonding?

_____ [1 mark]

b Look at the diagram.

Explain how a water molecule is formed from atoms of other elements.

_____

_____

_____ [4 marks]

c Carbon dioxide and water do not conduct electricity. Explain why.

_____ [1 mark]

d The formation of simple molecules containing single and double covalent bonds can be represented by dot and cross models.

i Draw the dot and cross diagram of water.

[2 marks]

ii Draw the dot and cross diagram of carbon dioxide.

[2 marks]

2 a Carbon dioxide and water have very low melting points. Use the idea of **intermolecular forces** to explain why.

_____

_____ [2 marks]

3 a Sodium is in group 1. Explain why.

_____ [1 mark]

b Chlorine atoms have 7 electrons in the outer shell. In which group is chlorine?

_____ [1 mark]

c i To which period does fluorine belong?

_____ [1 mark]

ii Explain why.

_____ [1 mark]

d The electronic structure of sulphur is 2, 8, 6. In which group is sulphur?

_____ [1 mark]

The group 1 elements

1 a Lithium, sodium and potassium react with water.

   i Which gas is given off?

_____ [1 mark]

   ii They float on the surface. Explain why.

_____ [1 mark]

b Sodium reacts very vigorously with water and forms sodium hydroxide.
Write down the word equation for the reaction of sodium with water.

_____ [2 marks]

c Reactivity of the alkali metals with water increases down group 1.

reactivity increases down		melting point in °C	boiling point in °C
	$_3$Li	179	1317
	$_{11}$Na	98	
	$_{19}$K		774

Estimate the melting point of potassium _____ and

the boiling point of sodium _____. [2 marks]

d Group 1 metals have similar properties. Explain why.

_____ [1 mark]

e Write a balanced symbol equation for the reaction of sodium metal with water.

_____ [2 marks]

2 a Marie and Mitch carried out some flame tests. How did they do this?

_____
_____
_____ [4 marks]

b Draw a **straight** line to match up their results.

red   potassium

yellow   lithium

lilac   sodium   [2 marks]

3 a Alkali metals have similar properties because when they react their atoms need to lose one electron to form full outer shells. Write down the equation to show the formation of a lithium ion from its atom.

_____ [1 mark]

b Sodium is more reactive than lithium. Explain why. Use ideas about numbers of shells of electrons.

_____
_____ [3 marks]

c What is the process of electron loss called?

_____ [1 mark]

The group 7 elements

D–C

1 a There is a **trend** in the **physical appearance** of the halogens at room temperature. Finish the table.

chlorine	
bromine	*orange liquid*
iodine	

[2 marks]

b Group 7 elements have similar properties. Explain why.

_____ [1 mark]

B–A*

c Chlorine has an electronic structure of 2, 8, 7. It gains one electron to become 2, 8, 8.

i Write an ionic equation to show this.

_____ [1 mark]

ii What is the process of electron gain called?

_____ [1 mark]

iii Fluorine is more reactive than chlorine. Explain why. Use ideas about the gain of electrons.

_____

_____ [2 marks]

D–C

2 a When a halogen reacts with an alkali metal a **metal halide** is made. Write down the word equation for the reaction between potassium and iodine.

_____ [2 marks]

B–A*

b Potassium reacts with chlorine to produce potassium chloride. Write a balanced equation to show this reaction.

_____ [2 marks]

D–C

3 a If halogens are bubbled through **solutions of metal halides** there are two possibilities: **no reaction**, or a **displacement reaction**.

i If chlorine is bubbled through potassium bromide solution a red-brown colour is seen. Explain why.

_____ [1 mark]

ii If bromine is bubbled through potassium chloride solution there is no reaction. Explain why.

_____ [1 mark]

b Bromine displaces iodine from potassium iodide solution.

i Write down a word equation for this reaction.

_____ [2 marks]

ii Write down a balanced equation for this reaction.

_____ [2 marks]

Electrolysis

1 a Explain the key features of the electrolysis of dilute sulphuric acid.

_____

_____

_____

_____

_____

_____ [6 marks]

b Explain why the volume of hydrogen gas and the volume of oxygen gas given off in this process are different.

_____ [1 mark]

D–C

2 The electrolysis of sodium chloride takes place in solution.
Describe the reactions at each of the electrodes. Explain the electron transfer processes in each case.

a At the cathode _____

_____

_____ [3 marks]

b At the anode _____

_____

_____ [3 marks]

B–A*

3 a Write about the key features of the production of aluminium by electrolytic decomposition.

_____

_____

_____

_____ [4 marks]

b Write down the word equation for the decomposition of aluminium oxide.

_____ [1 mark]

c The electrode reactions in the electrolytic extraction of aluminium involve the transfer of electrons. Explain how reduction and oxidation occur at the electrodes.

_____

_____

_____

_____ [4 marks]

d Why is the chemical cryolite added to aluminium oxide?

_____

_____ [2 marks]

D–C

B–A*

Transition elements

Grades

D–C

1 a A compound that contains a transition element is often coloured.

   i What is the colour of copper compounds?

_____

   ii What is the colour of iron(II) compounds?

_____

   iii What is the colour of iron(III) compounds?

_____ [3 marks]

b A transition metal and its compounds are often catalysts.

   i Which transition metal is used in the Haber process to produce ammonia?

_____ [1 mark]

   ii If the metal used to harden margarine is number 28, suggest whether this is a transition metal or not. Use the periodic table on page 112 to help you.

_____ [1 mark]

c If a transition metal carbonate is heated, it decomposes to form a metal oxide and carbon dioxide. Write down a word equation for the decomposition of copper carbonate.

_____ [1 mark]

D–C

2 a Sodium hydroxide solution is used to identify the presence of transition metal ions in solution. Finish the table.

ion	colour
Cu^{2+}	
Fe^{2+}	
Fe^{3+}	

[3 marks]

B–A*

3 a Copper carbonate decomposes into copper oxide and carbon dioxide. Write the balanced symbol equation for this thermal decomposition.

_____ [2 marks]

b Iron(III) ions react with hydroxide ions to from a precipitate. Write a balanced ionic symbol equation to show this precipitation reaction.

_____ [3 marks]

Metal structure and properties

1 a Silver is often chosen to make a piece of jewellery. Put a (ring) around the **two** properties that are important for this.

D–C

ductile   good electrical conductor   high boiling point

high melting point   lustrous   malleable

good thermal conductivity   [2 marks]

b Copper is often used for the base or the whole of a saucepan. Use your knowledge about chemical properties to explain why.

_____

_____

_____   [3 marks]

B–A*

c Aluminium is used in the aircraft industry and in modern cars. Explain why.

_____   [1 mark]

d Label the diagram to explain a **metallic bond**.

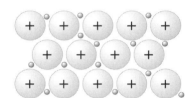

[1 mark]

e Metals have high melting points and boiling points. Use ideas about delocalised electrons to explain why.

_____

_____   [3 marks]

2 a What are superconductors?

D–C

_____   [1 mark]

b What are **two** potential benefits of superconductors?

_____

_____   [2 marks]

c Why do scientists need to develop superconductors that will work at 20 °C?

B–A*

_____   [1 mark]

d Explain why metals conduct electricity.

_____   [1 mark]

e

metal	electrical conductivity 10^6/cm Ω	thermal conductivity W m^{-1} K^{-1}	tensile strength MPa
A	0.099	80	350
B	0.452	320	170
C	0.596	400	210

i Which metal would be used for electrical cables? _____ [1 mark]

ii Which metal would be used for pulling up elevators (lifts)? _____ [1 mark]

C3 Revision checklist

- I know that the nucleus is made up of protons and neutrons, with each having a relative mass of 1. ☐

- I know that electrons surround the nucleus and occupy shells in order. They have almost no mass, 0. ☐

- I know that positive ions are formed by the loss of electrons from the outer shell. ☐

- I know that negative ions are formed by the gain of electrons into the outer shell. ☐

- I know that non-metals combine by sharing electrons, which is called covalent bonding. ☐

- I can draw the 'dot and cross' diagrams of simple molecules such as H_2, Cl_2, CO_2 and H_2O. ☐

- I know that lithium, sodium and potassium react vigorously with water and give off hydrogen. ☐

- I know that group 1 metals have one electron in their outer shell, which is why they are similar. ☐

- I know that chlorine is a green gas, bromine is an orange liquid and iodine is a grey solid. ☐

- I know that halogens gain one electron to form a stable outer shell. This is called reduction. ☐

- I know that in the electrolysis of dilute sulphuric acid, H_2 is made at the cathode and O_2 at the anode. ☐

- I know that when aluminium oxide is electrolysed, Al is formed at the cathode and O_2 at the anode. ☐

- I know that compounds of copper are blue, iron(II) are light green and iron(III) are orange/brown. ☐

- I know that metals conduct electricity as the 'sea' of electrons move through the positive metal ions. ☐

Speed

1 Freddie is on holiday in France. He travels 390 km in 3 hours on a French motorway.

a Calculate the average speed of his car on this journey in km/h.

_____

_____ [3 marks]

b Why is your answer the **average** speed of the car?

_____ [1 mark]

c The speed limit on the motorway is 130 km/h. Did the car break the speed limit? Explain your answer.

_____

_____ [2 marks]

D–C

2 a A cycle track is 500 m long. Imran completes 10 laps (he rides round the track 10 times). Imran travelled at an average speed of 12.5 m/s.

i How long did he take to complete ten laps?

_____

_____ [5 marks]

ii Imran put on a spurt in the last lap, completing it in 35 s. What was his average speed, in m/s, for the last lap?

_____

_____ [3 marks]

iii Calculate Imran's average speed for the first nine laps.

_____

_____

_____ [4 marks]

B–A*

b The graph shows Ashna's walk to her local shop and home again.

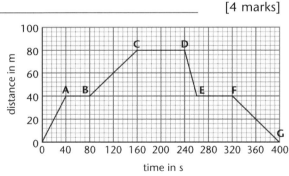

i During which part of her journey did she walk fastest?

_____ [1 mark]

ii Calculate Ashna's speed between 0 and A.

_____

_____ [3 marks]

iii Calculate Ashna's speed between F and G.

_____

_____ [2 marks]

D–C

B–A*

Changing speed

1 Darren is riding his bicycle along a road. The speed-time graph shows how his speed changed during the first minute of his journey.

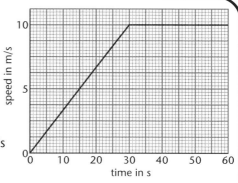

a Darren makes the same journey the next day but:
- increases his speed at a steady rate for the first 20 s, reaching a speed of 10 m/s
- travels at a constant speed for 10 s
- slows down at a steady rate for 15 s to a speed of 5 m/s
- travels at a constant speed of 5 m/s.

Plot the graph of this journey on the axes given.

[4 marks]

b How could you calculate the distance Darren travelled in the first minute of his original journey?

_____ [1 mark]

c The way in which the speed of a car changes over a 60 s period is shown in the table.

i Plot a speed-time graph for the car using the axes given. [5 marks]

time in s	speed in m/s
0	0
5	5
10	10
15	15
20	15
25	15
30	15
35	15
40	15
45	11
50	7.5
55	3.5
60	0

ii The car is in an area where the speed limit is 50 km/h. Does the driver exceed this limit? Show clearly how you decided.

_____
_____ [4 marks]

iii Use the graph to calculate the acceleration of the car.

_____
_____ [3 marks]

iv Use the graph to calculate the distance travelled by the car.

_____
_____ [4 marks]

2 a A car accelerates from 10 to 40 m/s in 6 s. Calculate its acceleration.

_____
_____ [4 marks]

b A cheetah accelerates at 6 m/s^2. How fast is it moving after 5 seconds, starting from rest?

_____
_____ [4 marks]

Forces and motion

1 A car of mass 500 kg accelerates steadily from 0 to 40 m/s in 20 s.

a What is its acceleration?

_____

_____

_____ [4 marks]

b What resultant force produces this acceleration?

_____

_____ [3 marks]

c The actual force required will be greater than your answer to **b**. Why?

_____ [1 mark]

d Calculate the acceleration of the car if the resultant force is increased to 1250 N.

_____

_____ [3 marks]

e The Moon orbits Earth due to gravitational attraction.

i Add an arrow to Earth, labelled F_E, to show the force on Earth due to the Moon. [1 mark]

ii Add an arrow to the Moon, labelled F_M, to show the force on the Moon due to Earth.

Moon [1 mark]

Earth

iii Write down **three** things about this pair of forces.

_____

_____

_____ [3 marks]

D–C

B–A*

B–A*

2 Helen is driving her car on a busy road when the car in front brakes suddenly. She puts her foot firmly on her brake pedal and just manages to stop without hitting the car in front.

a Write down **two** things, apart from speed, that could increase Helen's thinking distance.

_____ [2 marks]

b Later that day Helen is driving at high speed on a motorway.

i How will Helen's thinking distance change?

_____ [1 mark]

ii Explain why.

_____ [2 marks]

c Why do worn tyres increase the braking distance?

_____

_____ [2 marks]

D–C

B–A*

Work and power

D–C

1 a Calculate the amount of work Hilary does when she lifts a parcel of weight 80 N onto a shelf 2 m above the ground.

_____

_____ [3 marks]

B–A*

b How high could Hilary lift a parcel of weight 60 N if she did the same amount of work?

_____

_____ [3 marks]

c The brakes of a car produce a force of 6000 N.

i Calculate the braking distance if the car has to lose 300 000 J of kinetic energy.

_____

_____

_____ [4 marks]

ii What happens to the kinetic energy lost by the car in braking?

_____ [2 marks]

2 Chris and Abi both have a mass of 60 kg. They both run up a flight of stairs 3 m high. Chris takes 8 s and Abi takes 12 s.

D–C

a Calculate Chris' weight. [Take g = 10 N/kg.]

_____ [2 marks]

b How much work does Chris do in running up the stairs?

_____

_____ [3 marks]

c Calculate Chris' power.

_____

_____ [3 marks]

B–A*

d Priya produces the same power as Abi but she runs up the stairs in 10 s. Calculate Priya's mass.

_____

_____

_____

_____ [5 marks]

D–C

3 Why should we keep our fuel consumption to a minimum to protect the environment?

_____

_____ [3 marks]

Energy on the move

P3 FORCES FOR TRANSPORT

1 Use the data about fuel consumption to answer the following questions.

car	fuel	engine size in litres	miles per gallon	
			urban	non-urban
Renault Megane	petrol	2.0	25	32
Land Rover	petrol	4.2	14	24

a Why is the fuel consumption better in non-urban conditions?

_____

_____ [2 marks]

b How many gallons of petrol would a Land Rover use on a non-urban journey of 96 miles?

_____ [2 marks]

c Would a Renault Megane use more or less fuel for the same journey? _____ [1 mark]

d Write down a reason for the difference.

_____ [1 mark]

D–C

2 a Which possesses more kinetic energy?

	mass in kg	speed in m/s		mass in kg	speed in m/s
A	2	5	B	2	7
C	20	2	D	15	2

A or B _____   C or D _____ [2 marks]

b Sam is driving a car of mass 1200 kg at a speed of 20 m/s.

i Calculate the kinetic energy of the car.

_____ [4 marks]

ii When Sam suddenly applies the brakes, the car travels 32 m before it stops. Calculate the braking force.

_____ [4 marks]

iii Sam suggests that the braking distance would be 16 m if he was travelling at 10 m/s. Do you agree? Explain your answer.

_____ [2 marks]

D–C

B–A*

3 a How do battery-powered cars cause pollution?

_____ [2 marks]

b Give **one** advantage of solar-powered cars compared to battery-powered cars.

_____ [1 mark]

c Give **one** disadvantage of solar-powered cars compared to battery-powered cars.

_____ [1 mark]

D–C

Crumple zones

1 Kevin was involved in an accident on the M1 motorway. Luckily, he was not seriously hurt, but his car was badly damaged.

a Finish the table to show how each feature helps to reduce Kevin's injuries by absorbing energy.

safety feature	how it works
seatbelt	
crumple zones	
air bag	

[3 marks]

b **i** What is meant by an 'active safety feature' on a car?

_____ [2 marks]

ii What is meant by a 'passive safety feature' on a car?

_____ [2 marks]

c Traction control is an active safety feature. Explain how it contributes to safety.

_____
_____ [2 marks]

d Compare the effect of active and passive safety features on road safety.

_____
_____ [2 marks]

e **i** What does **ABS** stand for?

_____ [1 mark]

ii Why are ABS brakes safer when a driver has to slam his foot on the brakes to stop quickly?

_____
_____ [2 marks]

2 To minimise injury the forces acting on the people in a car during a car accident must be as small as possible.

a Explain why this means safety features must reduce the deceleration of the car on impact.

_____
_____ [2 marks]

b Explain how **one** safety feature is designed to reduce the deceleration of the car on impact.

_____
_____ [3 marks]

Falling safely

1 a Charlie drops a golf ball and a ping pong ball from a height of 30 cm above a table.
Both balls hit the table together although their masses are different.
Charlie now drops the two balls from a height of 130 cm above the table.
Explain why the golf ball hit the table before the ping pong ball.

_____

_____ [2 marks]

b Sarah is a sky diver. She has a mass of 60 kg.

   i What is the value of her acceleration just
after leaving the aircraft?

_____ [1 mark]

c **i** On the diagram, mark and name the forces acting on Sarah as she falls. [2 marks]

   ii What can you say about the size of these forces?

_____ [1 mark]

d Sarah's acceleration decreases as she falls. Explain why.

_____ [2 marks]

e **i** Eventually she is travelling at a constant speed. What is this speed called?

_____ [1 mark]

   ii What can you say about the size of the forces acting on her now?

_____ [1 mark]

f **i** Sarah's brother weighs 1000 N. He also sky dives. Will he travel at a larger or
smaller constant speed? Explain your answer.

_____

_____

_____ [4 marks]

g Sarah opens her parachute.

   i What happens to each of the forces acting on her now?

_____

_____ [2 marks]

   ii Explain how Sarah is able to land safely.

_____

_____ [4 marks]

2 a What can you say about the forces acting on a car when it is travelling at a
constant speed?

_____ [1 mark]

b Suggest what may affect the maximum speed of a car.

_____ [2 marks]

c What do car designers do to make the maximum speed as large as possible?

_____ [2 marks]

Grades column markers: D–C, D–C, B–A*, B–A*

The energy of theme rides

D–C

1 Kate is bouncing a ball. She drops it from A and it rises to **D** after the first bounce.

a Why is **D** much lower than **A**?

_____

_____

[2 marks]

B–A*

b The ball has a mass of 40 g. **A** is 1.2 m above the ground. Calculate the gravitational potential energy of the ball at **A**.

_____

_____

[4 marks]

c Kate drops the ball from increasing heights. She finds there is a maximum height that the ball bounces to. Explain why.

_____

_____

[3 marks]

D–C

2 The diagram shows a roller coaster. The carriages are pulled up to **B** by a motor and then released.

a At which point, **A**, **B**, **C**, **D** or **E**, do the carriages have the greatest gravitational potential energy?

_____ [1 mark]

b At which point, **A**, **B**, **C**, **D** or **E** do the carriages have the greatest kinetic energy?

_____

[1 mark]

c Describe the main energy change as the carriages move from **B** towards **C**.

_____

[2 marks]

d Why must the height of the next peak at **D** be less than that at **B**?

_____

_____

[2 marks]

e The theme park decides to build a faster roller coaster. Suggest how they could modify the design to achieve this, using your knowledge of energy transfers.

_____

_____

_____

[3 marks]

B–A*

3 a Penny has a mass of 65 kg. What is her weight? [Take g = 10 N/kg.]

_____

[2 marks]

b Penny travels to another planet where she only weighs 400 N. Calculate the gravitational field strength on the planet.

_____

_____

[3 marks]

P3 Revision checklist

- I can state and use the formula: speed = distance ÷ time, including a change of subject. ☐

- I can draw and interpret distance-time graphs and speed-time graphs. ☐

- I can use the formula: acceleration = change in speed ÷ time taken, including a change of subject. ☐

- I can state and use the formula: force = mass x acceleration, including a change of subject. ☐

- I can describe the factors that may affect thinking, braking and stopping distances. ☐

- I can state and use the formula: work done = force x distance, including a change of subject. ☐

- I can state and use the formula: power = work done ÷ time, including a change of subject. ☐

- I can use the formula: $KE = \frac{1}{2}mv^2$ and apply it to braking distances and other examples. ☐

- I can interpret data about fuel consumption and explain the factors that affect these figures. ☐

- I can describe and evaluate the effectiveness of various car safety features. ☐

- I can describe how and use forces to explain why falling objects may reach a terminal speed. ☐

- I can recognise that acceleration in free fall (g) is constant. ☐

- I can interpret a gravity ride (roller coaster) in terms of PE, KE and energy transfer. ☐

- I can state and use the formula: weight = mass x g, including a change of subject. ☐

Who planted that there?

D–C

1 Look at the diagram. It shows the inside of a leaf.

   a Label

      i the upper epidermis

      ii a palisade cell

      iii a mesophyll cell

   b Put an **S** to show the
position of a stoma.   [3 marks]

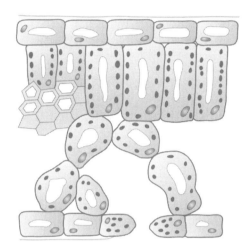

[1 mark]

D–C

2 Finish this sentence about gas exchange.
Carbon dioxide enters the leaf and oxygen leaves by _____ .   [1 mark]

D–C

3 a A leaf is adapted for photosynthesis. It has broad leaves. This gives a large surface
area to absorb light. Write down **three other** ways they are adapted for photosynthesis.

    _____

    _____

    _____   [3 marks]

B–A*

   b The cells of a leaf are also adapted for maximum photosynthesis.
Explain how the cells are adapted.

     i Palisade cells _____

     _____

     _____   [2 marks]

     ii Mesophyll cells _____

     _____

     _____   [2 marks]

Water, water everywhere

1 a Lauren investigates osmosis. She put some potato chips into pure water and some into salt water. She left them for two hours.
The diagram shows her results. Explain why the potato chip went floppy in salt water.

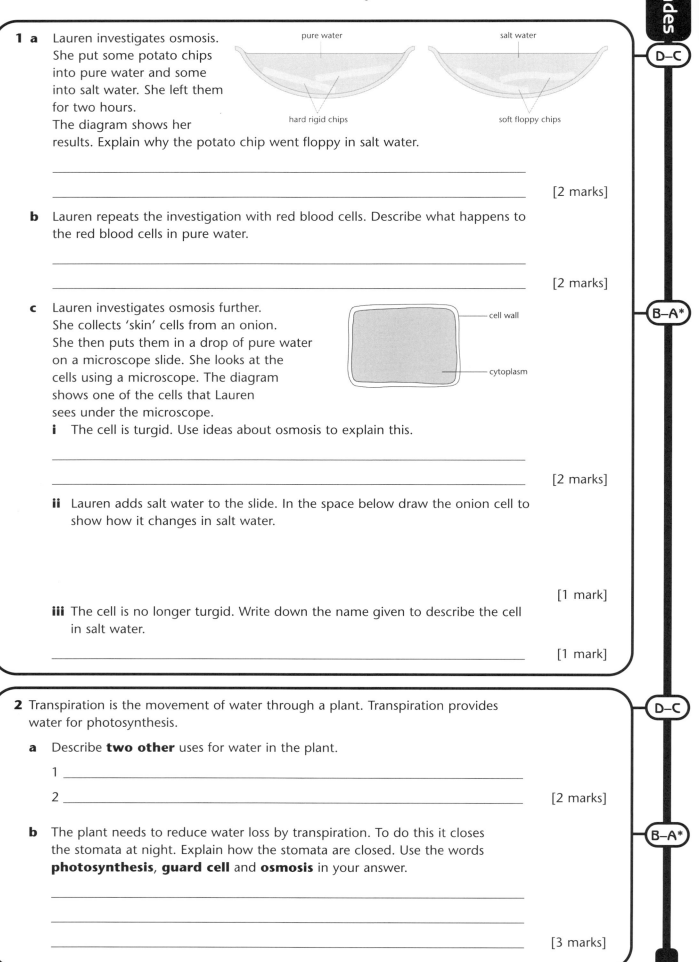

pure water   salt water

hard rigid chips   soft floppy chips

_____

_____ [2 marks]

b Lauren repeats the investigation with red blood cells. Describe what happens to the red blood cells in pure water.

_____

_____ [2 marks]

c Lauren investigates osmosis further. She collects 'skin' cells from an onion. She then puts them in a drop of pure water on a microscope slide. She looks at the cells using a microscope. The diagram shows one of the cells that Lauren sees under the microscope.

cell wall

cytoplasm

i The cell is turgid. Use ideas about osmosis to explain this.

_____

_____ [2 marks]

ii Lauren adds salt water to the slide. In the space below draw the onion cell to show how it changes in salt water.

[1 mark]

iii The cell is no longer turgid. Write down the name given to describe the cell in salt water.

_____ [1 mark]

2 Transpiration is the movement of water through a plant. Transpiration provides water for photosynthesis.

a Describe **two other** uses for water in the plant.

1 _____

2 _____ [2 marks]

b The plant needs to reduce water loss by transpiration. To do this it closes the stomata at night. Explain how the stomata are closed. Use the words **photosynthesis**, **guard cell** and **osmosis** in your answer.

_____

_____

_____ [3 marks]

D–C

B–A*

D–C

B–A*

Transport in plants

D–C

1 Look at the diagram. It shows the arrangement of cells in a cross section of a root.

A   B

a Write down the names of the **two** types of cells shown in the diagram.

Cell **A** _____

Cell **B** _____

[2 marks]

b Describe the function of cell **B**.

_____

_____

[2 marks]

B–A* **c** Explain how the structure of **B** helps it carry out its function.

_____

_____

[2 marks]

D–C

2 Jasmine investigates transpiration in plants. She sets up two lines using this apparatus.

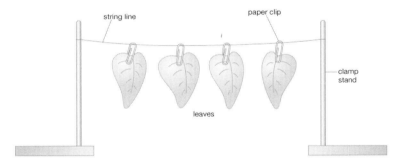

string line   paper clip

clamp stand

leaves

a One line is kept in the dark and one line is kept in the light. The leaves in the **light** lost 7.4 g in mass. Suggest how much mass the leaves in the **dark** lost.

_____ [1 mark]

B–A* **b** Jasmine set up a third line and placed a clear plastic bag over the leaves.
i Describe how this will affect the rate of transpiration.

_____ [1 mark]

ii Explain how the plastic bag changed the rate of transpiration.

_____

_____

[2 marks]

Plants need minerals too

1 a Finish the table to show why a plant needs certain minerals.
The first one has been done for you.

mineral	why the mineral is needed	how the mineral is used
nitrate	*for cell growth*	*nitrogen is used to make amino acids*
phosphate		
potassium		
magnesium		

[6 marks]

D–C

b Mineral deficiency causes poor plant growth. A plant grown without nitrates will have poor growth and yellow leaves. Describe how a plant will look if it is grown without

   i phosphate _____

   _____ [2 marks]

   ii potassium _____

   _____ [2 marks]

   iii magnesium _____

   _____ [1 mark]

B–A*

2 Look at the graph. It shows the uptake of minerals by algae living in pond water.

D–C

   a What is the concentration of chlorine
     i in the cells of the algae?

     _____

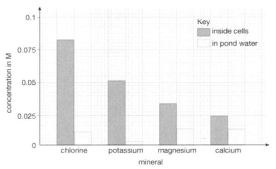

[1 mark]

     ii in the pond water?

     _____ [1 mark]

   b Chlorine is not absorbed by diffusion. Explain why.

     _____

     _____ [2 marks]

B–A*

   c Name the type of transport that the algae use to take up chlorine from the pond water.

     _____ [1 mark]

   d Destroying the mitochondria inside the algae stops the uptake of minerals. Explain why.

     _____

     _____ [2 marks]

Energy flow

D–C

1 Look at the diagram. It shows a pyramid of numbers.

10s of foxes

100s of shrews

1000s of caterpillars

oak tree

   a Explain what information is shown in a pyramid of numbers.

   _____

   _____

   _____

   [2 marks]

   b In the space below draw a pyramid of biomass for the same food chain.

   [1 mark]

   c As energy flows through the food chain some is lost. Write down **one** way in which energy is lost.

   _____

   [1 mark]

B–A*

2 Farmers who grow crops produce large amounts of biomass. The biomass has many uses. They may choose to eat the biomass or feed it to their cows.

   a Suggest **two other** ways they may choose to use the biomass.

   1 _____

   2 _____   [2 marks]

   b Look at the diagram. It shows the energy transfer from the farmer's crops to the cow.

   1022 kJ in heat loss

   Sun

   energy for growth

   1909 kJ in waste

   3090 kJ energy in 1 m² of crops

   i Calculate the amount of energy used for growth.

   Energy used for growth _____kJ   [2 marks]

   ii Calculate the efficiency of energy transfer of the cow.

   Energy efficiency _____%   [2 marks]

   c Some people think it is better to grow crops for food instead of producing meat to eat.
   Use your answer to part **b** to explain why.

   _____

   [1 mark]

D–C

3 a Brazil produces a lot of sugar cane. The sugar cane is used to produce a biofuel. Describe how. Use the words **yeast**, **fermentation** and **petrol** in your answer.

   _____

   _____

   _____

   [3 marks]

B–A*

   b Describe **one** advantage and **one** disadvantage of using biofuels.

   Advantage _____

   _____

   [1 mark]

   Disadvantage _____

   _____

   [1 mark]

Farming

1 a DDT is a chemical that can be used to kill insects.
Look at the food chain from a marine environment.

plankton ⟶ krill ⟶ penguin ⟶ seal

DDT contaminated the water. The plankton absorbed the DDT but did not die.
The seals began to die from the DDT.
Explain why.

_____

_____

_____ [3 marks]

D–C

b Look at the picture. It shows intensive farming of pigs.
This type of farming is more energy efficient than keeping
pigs outside. Explain why.

_____

_____

_____

[2 marks]

B–A*

2 a Describe how tomatoes can be grown without soil.

_____

_____ [2 marks]

D–C

b Write down **one** advantage and **one** disadvantage of growing tomatoes without soil.

Advantage _____

_____

Disadvantage _____

_____ [2 marks]

B–A*

3 a Organic farmers use a method of farming called crop rotation to help the soil.
Describe **two other** methods of farming used by organic farmers.

1 _____ [1 mark]

2 _____ [1 mark]

D–C

b Organic farming is not possible in countries such as Ethiopia.
Suggest why.

_____

_____ [2 marks]

B–A*

c In 1935 large cane toads were introduced into Australia to control insects feeding on sugar
cane. The toads did not eat the insects; they ate native toads instead. The cane toads have
no predators in Australia.

i Write down the name given to this type of pest control.

_____ [1 mark]

D–C

ii The cane toads are now bigger pests than the insects they were sent to control.
Suggest **one** reason why.

_____ [1 mark]

Decay

D–C

1 a Look at the apparatus Shahid uses to investigate decay. Both samples are weighed, left for two days and then re-weighed.

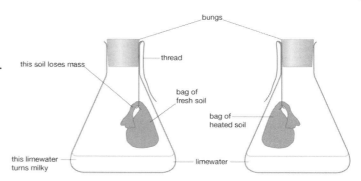

bungs

thread

this soil loses mass

bag of fresh soil

bag of heated soil

this limewater turns milky

limewater

The table shows Shahid's results.

mass in g	bag of fresh soil	bag of heated soil
at start	6.2	6.5
at end	5.8	6.5
change		0

i Calculate the change in mass of the bag of fresh soil. Write your answer in the shaded box. [1 mark]

ii The heated soil did not change in mass. Explain why.

_____

_____ [2 marks]

b Earthworms feed on the remains of dead and decaying organisms. How do earthworms help microorganisms to increase the rate of decay?

_____

_____ [2 marks]

B–A*

c Shahid extends the investigation. He sets up four more bags of fresh soil and leaves them at different temperatures.

i Predict which of Shahid's bags lost the most mass. Put a ring around the correct answer.

bag A 10 °C   **bag B 20 °C**   **bag C 40 °C**   **bag D 60 °C** [1 mark]

ii Explain your answer to part **c i**.

_____

_____ [2 marks]

d Fungi such as mushrooms are called saprophytes. Describe how fungi feed on dead plant material.

_____

_____ [2 marks]

D–C

2 Jennifer grows strawberries but she has too many to eat before they decay. She preserves the strawberries by making jam. Explain how this method slows down decay of the strawberries.

_____

_____ [2 marks]

Recycling

1 The diagram shows the carbon cycle.

a Finish labelling the carbon cycle. Choose **three** words from this list.

> **burning**
> **digestion**
> **feeding**
> **photosynthesis**
> **reproduction**
> **respiration**

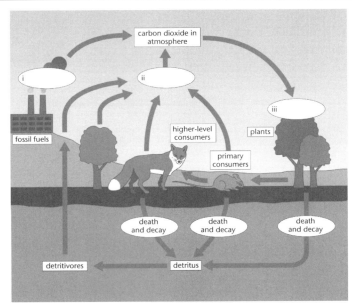

D–C

[3 marks]

b Write down the name of the process that removes carbon dioxide from the air.

_____ [1 mark]

c Decomposers return carbon dioxide to the air. Explain why.

_____

_____ [2 marks]

d Look at the picture. It shows marine animals called corals. The carbon in the shells of coral is recycled over millions of years. Explain how this occurs.

B–A*

_____

_____

_____

_____ [3 marks]

---

2 In the nitrogen cycle, plants take up nitrates from the soil.

D–C

a Why do plants need nitrates?

_____ [1 mark]

b The nitrates are returned to the soil by decomposers. Explain how.

_____

_____ [2 marks]

c Nitrogen gas in the air cannot be used directly by plants. Explain why.

B–A*

_____ [1 mark]

d Describe **two** ways nitrogen gas can be turned into nitrates for the plants.

1 _____

2 _____ [2 marks]

e Write down the name of the bacteria that return nitrogen to the air.

_____ [1 mark]

B4 Revision checklist

- I can label a diagram showing the parts of a leaf. ☐

- I can explain how a leaf is adapted for photosynthesis. ☐

- I know that osmosis is the movement of water molecules across a partially-permeable membrane. ☐

- I can describe the structure of xylem and phloem. ☐

- I can explain how transpiration rate can be increased. ☐

- I know that active transport needs energy from respiration. ☐

- I can identify mineral deficiencies in plants. ☐

- I can construct pyramids of numbers and biomass. ☐

- I can explain the reasons for developing biofuels. ☐

- I can describe the difference between intensive farming and organic farming. ☐

- I know that a saprophyte decays organisms by releasing enzymes. ☐

- I can explain why the different preservation methods stop food decay. ☐

- I can describe the carbon cycle. ☐

- I can describe the nitrogen cycle. ☐

Acids and bases

1 a i What is an alkali?

_____ [1 mark]   D–C

ii Finish the word equation for neutralisation.

_____ + base → salt + _____ [2 marks]

b Write down the word equation for the reaction between copper carbonate and sulphuric acid.

_____ + _____ → _____ + _____ + _____   B–A*

[3 marks]

c Write down the name of the compound formed when sodium hydroxide reacts with nitric acid.

_____ [1 mark]

d Finish and balance the symbol equations for the reactions between

i $HCl + NaOH \rightarrow$ _____ $+ H_2O$ [1 mark]

ii hydrochloric acid and calcium carbonate $CaCO_3$

_____ [3 marks]

---

2 a Which ion is found in an acid solution?   B–A*

_____ [1 mark]

b Which ion is found in an alkaline solution?

_____ [1 mark]

c If these **two** types of ion react together, what is made?

_____ [1 mark]

---

3 a How does the pH of an acid change when an alkali is added to it?   D–C

_____ [1 mark]

b **Universal indicator solution** can be used to measure the acidity of a solution. A few drops are added to the test solution and then the colour of the solution is compared to a standard colour chart.
Describe how the colour changes when a strong acid is added to an alkali to neutralise it.

_____

_____

_____ [3 marks]

Reacting masses

D–C

1 a Work out the relative formula mass of NaOH.

_____ [1 mark]

b Work out the relative formula mass of $CaCO_3$.

_____ [1 mark]

c Work out the relative formula mass of $Ca(OH)_2$.
Use the relative atomic masses.

H 1   C 12   O 16   Na 23   Ca 40

B–A*

_____ [1 mark]

d Explain why mass is conserved in a chemical reaction.

_____

_____ [2 marks]

D–C

2 a Leo and Lesley made some crystals of magnesium sulphate. They did not make as much as they hoped. They wanted to make 42 g. They only made 28 g.
  i What was their 'actual yield'?

_____ [1 mark]

  ii What was their 'predicted yield'?

_____ [1 mark]

  iii How will they calculate their percentage yield?

_____

_____ [1 mark]

  iv What was their percentage yield?

_____

_____ [2 marks]

B–A*

b Zinc carbonate decomposes on heating to give zinc oxide and carbon dioxide.
  i Write a balanced symbol equation for this reaction.

_____ [2 marks]

  ii Calculate how much CO_2 is made when 12.50 g of $ZnCO_3$ decomposes.

_____

_____

_____

_____ [4 marks]

Fertilisers and crop yield

1 a Why do farmers add fertilisers to their crops?

_____ [1 mark]   D–C

b How do fertilisers get into the plants?

_____ [1 mark]

c Why is nitrogen needed for increased plant growth?   B–A*

_____ [1 mark]

2 a To calculate the yield when making a fertiliser you need to calculate its **relative formula mass**. What is the relative formula mass of ammonium sulphate $(NH_4)_2SO_4$? Use the periodic table on page 112.   D–C

_____ [2 marks]

b Farmers can use relative formula masses to find the percentage of each element in a fertiliser – it is printed on the bag for them. They use the formula: B–A*

$$\text{percentage of element} = \frac{\text{mass of the element in the formula}}{\text{relative formula mass}} \times 100$$

Calculate the percentage of nitrogen in potassium nitrate, KNO_3.

_____
_____
_____ [3 marks]

3 a Many fertilisers are **salts**, so they can be made by reacting acids with bases. What else is made?   D–C

$$\text{acid + base} \rightarrow \text{salt} + \underline{\hspace{3cm}}$$ [1 mark]

b Don and Demi want to make some ammonium phosphate.

i Which acid will they need to use?_____ [1 mark]

ii Which alkali will they need to use? _____ [1 mark]

iii Write down a word equation to show this reaction.

_____ [2 marks]

c Describe in steps how Don and Demi make ammonium phosphate from their acid and alkali.   B–A*

_____
_____
_____
_____ [6 marks]

4 Too much fertilisers applied may increase the level of nitrates or phosphates in a water course. This may cause eutrophication to occur. Explain how this happens.   B–A*

_____
_____
_____ [3 marks]

The Haber process

D–C

1 a Write about how ammonia is made. Include the conditions needed in your answer.

_____

_____

_____ [3 marks]

B–A*

b As the reaction for producing ammonia is reversible, the percentage yield for the reaction cannot be 100%.

i How does the pressure affect the reaction?

_____

_____ [2 marks]

ii How does the temperature affect the reaction?

_____

_____ [2 marks]

iii Give the optimum temperature for this reaction and explain why it is used.

_____

_____

_____ [2 marks]

D–C

3 a Look at the graph. At which pressure is most ammonia made at 400 °C ?

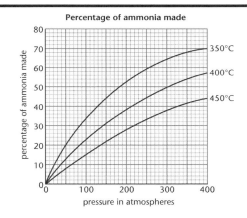

Percentage of ammonia made

percentage of ammonia made

350°C
400°C
450°C

pressure in atmospheres

_____ [1 mark]

b As the pressure increases what happens to the yield of ammonia?

_____ [1 mark]

c As the temperature increases what happens to the yield of ammonia?

_____ [1 mark]

B–A*

d Chemical plants need to work at the conditions that produce the highest percentage yield for a reaction, most cheaply. Use ideas about optimum conditions to explain how this is achieved in the manufacture of ammonia. Explain why these cut costs.

_____

_____

_____

_____ [5 marks]

Detergents

1 a A detergent can be made by **neutralising** an organic acid using an alkali. Write a word equation for this reaction.

_____ [2 marks]

b Why are detergents used to clean greasy plates?

_____

_____ [2 marks]

c New washing powders allow clothes to be washed at low temperatures.
  i This is good for the environment. Explain why.

_____

_____ [3 marks]

  ii It is also good for coloured clothes to be washed at low temperatures. Explain why.

_____

_____ [2 marks]

d When clothes are washed, grease is lifted off the clothes and put into water by detergent.
Use the words **hydrophilic** and **hydrophobic** to explain the stages in the diagrams.

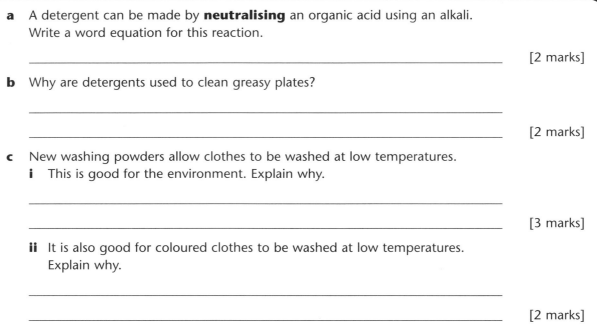

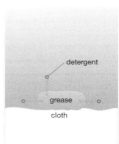 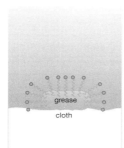 

[3 marks]

2 a The forces that hold molecules of grease together and molecules of dry-cleaning solvent together are forces between molecules. What are these are called?

_____ [1 mark]

b Molecules of water are held together by stronger forces called **hydrogen bonds**. The water molecules cannot stick to the grease because they are sticking to each other much too strongly. On the diagram mark the **covalent** bonds between atoms and the **hydrogen** bonds between water molecules.

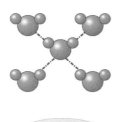 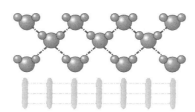

water sticks strongly to water

water sticks to water too strongly to stick to grease

[5 marks]

Batch or continuous?

D–C

1 a Why are pharmaceuticals made in small batches?

_____ [1 mark]

b How is the large scale production of ammonia different from the small scale production of pharmaceuticals?

_____ [1 mark]

B–A*

c A **continuous** process plant is effective. Explain why.

_____

_____

_____

_____ [4 marks]

d **Batch** processes are not as efficient but have one major advantage. Explain what this is.

_____ [1 mark]

D–C

2 a Draw **straight** lines to match the reasons for the high costs of making and developing medicine and pharmaceutical drugs to the **best** explanation.

| strict safety laws | The medicines are made by a batch process so less automation can be used. |

| research and development | They may be rare and costly. |

| raw materials | They take years to develop. |

| labour intensive | People need to feel a benefit without too many side effects. |

[3 marks]

B–A*

3 Whether or not a drug is developed depends on a number of economic considerations. Explain what these are.

_____

_____

_____

_____

_____ [6 marks]

Nanochemistry

1 a Draw a **straight** line to match the carbon to its correct structure.

diamond

graphite

buckminster
fullerene

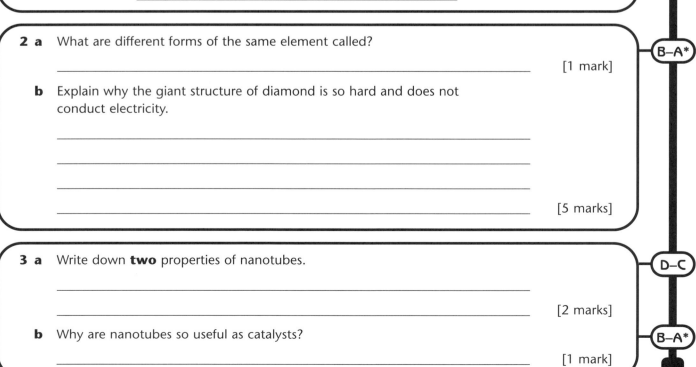

D–C

[3 marks]

b Finish the table to show **one** use of the carbon type and a reason for that use.

	diamond	graphite	buckminster fullerene
use			*semiconductors in electrical circuits good electrical*
reason			*conductor on small scale*

[4 marks]

c Explain why these structures have these properties.

diamond	graphite
Does not conduct electricity because _____ _____ _____	Slippery because _____ _____ _____

B–A*

[2 marks]

2 a What are different forms of the same element called?

_____

B–A*

[1 mark]

b Explain why the giant structure of diamond is so hard and does not conduct electricity.

_____

_____

_____

_____

[5 marks]

3 a Write down **two** properties of nanotubes.

_____

_____

D–C

[2 marks]

b Why are nanotubes so useful as catalysts?

_____

B–A*

[1 mark]

How pure is our water?

D–C

1 a The water in a river is cloudy and often not fit to drink. To make clean drinking water it is passed through a **water purification** works.
Label the **three** main parts of this process.

$$\boxed{\underline{\hspace{3cm}}} \rightarrow \boxed{\underline{\hspace{3cm}}} \rightarrow \boxed{\underline{\hspace{3cm}}}$$

[3 marks]

b Explain the stages of water purification in detail.

B–A*

_____

_____

_____

_____

_____

[3 marks]

c i Seawater has many substances dissolved in it so it is undrinkable. Which special technique has to be used to remove these unwanted substances?

_____

[1 mark]

ii Why is this technique not used for all purification of water?

_____

_____

[1 mark]

D–C

2 a Explain why relief organisations concentrate on providing clean water supplies.

_____

_____

[2 marks]

b Water is a **renewable resource** but not an endless resource in any one country. Explain why.

_____

_____

[2 marks]

D–C

3 a Write a word equation for the precipitation reaction between lead nitrate and potassium chloride.

_____

[2 marks]

B–A*

b Write a word equation for the precipitation reaction between silver nitrate $AgNO_{3(aq)}$ and potassium bromide $KBr_{(aq)}$.

_____

[1 mark]

c Write a full balanced symbol equation for this precipitation reaction.

_____

[3 marks]

C4 Revision checklist

- I know that neutralisation is a reaction where: acid + base → salt + water.

- I can construct balanced symbol equations, such as:
$2KOH + H_2SO_4 \rightarrow K_2SO_4 + H_2O$

- I can work out the relative formula mass of a substance from its formula, such as $Ca(OH)_2$.

- I can work out the percentage yield using the formula:
% yield = actual yield × 100 ÷ predicted yield.

- I know that fertilisers provide extra essential elements but excess can cause eutrophication.

- I know that ammonia is made by the Haber process where N_2 and H_2 are put over an iron catalyst.

- I know that the higher the pressure, the higher the percentage yield of ammonia.

- I know that a catalyst will increase the rate of reaction but will not change the percentage yield.

- I know that a detergent has a hydrophilic head and a hydrophobic tail.

- I know that dry cleaning is a process used to clean clothes using a solvent that is not water.

- I know that a continuous process makes chemicals all the time but a batch process does not.

- I can recognise the three allotropes of carbon: diamond, graphite and buckminster fullerene.

- I can explain that graphite is slippery and is used as electrodes as it conducts electricity.

- I know that water purification includes filtration, sedimentation and chlorination.

Sparks!

D–C

1 a Sally stands on an insulating mat. She puts her hands on the dome of an uncharged Van de Graaff generator. The Van de Graaff generator is switched on and Sally's hair starts to stand on end.

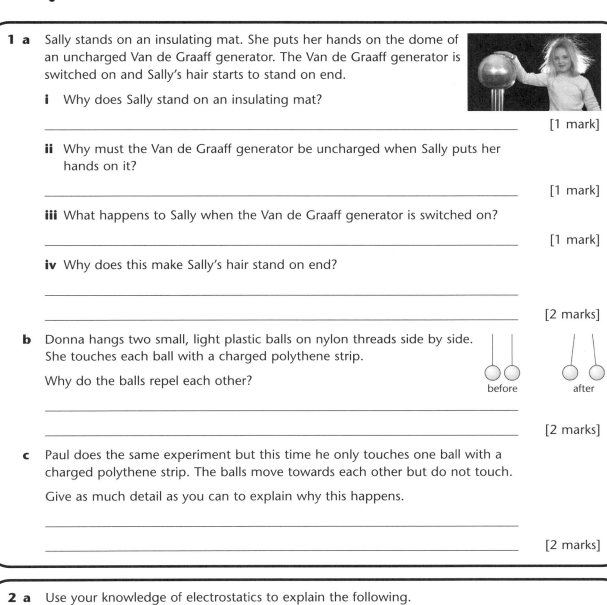

   i Why does Sally stand on an insulating mat?

   _____

      [1 mark]

   ii Why must the Van de Graaff generator be uncharged when Sally puts her hands on it?

   _____

      [1 mark]

   iii What happens to Sally when the Van de Graaff generator is switched on?

   _____

      [1 mark]

   iv Why does this make Sally's hair stand on end?

   _____

   _____

      [2 marks]

B–A*

b Donna hangs two small, light plastic balls on nylon threads side by side. She touches each ball with a charged polythene strip.

Why do the balls repel each other?

before   after

_____

_____

      [2 marks]

c Paul does the same experiment but this time he only touches one ball with a charged polythene strip. The balls move towards each other but do not touch.

Give as much detail as you can to explain why this happens.

_____

_____

      [2 marks]

D–C

2 a Use your knowledge of electrostatics to explain the following.

   i You sometimes get an electric shock on closing a car door after a journey.

   _____

   _____

      [2 marks]

   ii You should never shelter under a tree during a thunderstorm.

   _____

      [1 mark]

   iii Cling film often sticks to itself as it is unrolled.

   _____

      [1 mark]

B–A*

b A factory uses machinery with moving parts. How can the machinery become charged?

_____

      [1 mark]

c Why do the operators stand on rubber mats?

_____

_____

      [3 marks]

Uses of electrostatics

1 A defibrillator delivers an electric shock through the chest wall to the heart.

    a What does the electric shock do to the heart?_____

        [1 mark]

    b The paddles of a defibrillator, charged from a high voltage supply, are placed on the patient's chest. How does the operator ensure that there is good electrical contact with the patient's chest?

        _____

        _____ [2 marks]

    c A current of about 50 A passes through the patient for about 4 ms (0.004 s). In general, such a large current would be fatal. Why can it be used in this situation?

        _____ [1 mark]

    d A typical defibrillator has a power of about 100 kW. How much energy does the patient receive if the supply is switched on for 4 ms (0.004 s)?

        _____

        _____ [4 marks]

2 In a bicycle factory the frames are painted using an electrostatic sprayer. The paint is positively charged. The frames are given the opposite charge to the paint.

nozzle is charged up positively

object to be painted is negatively charged

    a Why does the paint spread out on leaving the sprayer?

        _____

        _____

        _____

        _____ [2 marks]

    b Why are the bicycle frames given the opposite charge to the paint?

        _____ [2 marks]

    c Freddie decides to repaint his bicycle frame. He uses a paint spray that charges the paint positively but he does **not** charge the frame.

        i What charge, if any, does Freddie's bicycle frame acquire? _____ [1 mark]

        ii What problem does this cause?

        _____ [1 mark]

3 An electrostatic precipitator is placed in the chimney of a power station. It contains some wires and plates which are connected to a high voltage supply.

    a If the wires are given a negative charge, what charge must be given to the plates?

        _____ [1 mark]

    b What charge do the soot particles gain as they pass close to the wires?

        _____ [1 mark]

Safe electricals

D–C

1 a Meera sets up the circuit shown.

   i What effect will this have on the brightness of the lamp?

_____ [1 mark]

variable
resistor

   ii Add an ammeter and voltmeter to Meera's circuit to allow her to measure the current through the lamp and the potential difference across the lamp. [2 marks]

   iii If the voltmeter reads 6 V and the ammeter 0.25 A, calculate the resistance of the lamp.

_____

_____ [4 marks]

B–A*

b Meera increases the potential difference across the lamp to 12 V. The current increases to 0.4 A.

   i What is the resistance now?

_____

_____ [2 marks]

   ii Explain the change in resistance in terms of electron movement.

_____

_____

_____ [4 marks]

D–C

2 A battery has positive and negative terminals. Give **two** differences between the voltage from a battery and mains voltage.

_____

_____ [2 marks]

D–C

3 a **i** Label the live (L), neutral (N) and earth (E) wires in the plug shown. [3 marks]

   ii Which wire, live, neutral or earth, is a safety wire?

_____ [1 mark]

   iii How does it work?

_____

_____ [2 marks]

B–A*

b **i** An electric kettle passes a current of 10.5 A when working normally. Should the plug contain a 5 A or 13 A fuse? _____ [1 mark]

   ii Why are fuses always connected in the live wire?

_____

_____ [2 marks]

   iii The kettle is made from metal. How do the fuse and earth wire stop a person receiving an electric shock if they touch the kettle when it is faulty?

_____

_____ [4 marks]

Ultrasound

1 a Sound is a longitudinal wave.

    i Explain how sound travels through the air to reach your ear.

_____

_____ [2 marks]

    ii How does the frequency of a note change if its pitch increases?

_____ [1 mark]

b What is 'ultrasound'?

_____ [1 mark]

c Light is an example of a transverse wave. What is the difference between a longitudinal and a transverse wave?

_____

_____ [2 marks]

D–C

B–A*

---

2 a Finish the sentences about an ultrasound scan. Choose words from this list.

echoes   gel   image   probe   pulse
reflected   skin   tissues   ultrasound

A _____ of ultrasound is sent into a patient's body. At each boundary between different

_____ some ultrasound is _____ and the rest is transmitted. The returning _____

are used to build up an _____ of the internal structure. A _____ is placed on the

patient's body between the ultrasound _____ and their _____. Without it nearly all

the _____ would be _____ by the _____ . [6 marks]

b Give **two** factors that affect the amount of ultrasound sent back to the detector at each interface within the body.

_____ [2 marks]

c Air has a density of 1.3 kg/m^3. Soft tissue has an average density of 1060 kg/m^3.

    i Explain why the gel used should have a density of about 1060 kg/m^3.

_____

_____ [2 marks]

    ii The time delay for an echo from ultrasound in soft tissue at a depth of 0.16 m was 0.2 ms (0.0002 s). Calculate the speed of ultrasound in soft tissue.

_____

_____ [4 marks]

D–C

B–A*

---

3 High-powered ultrasound is used to treat a patient with kidney stones.

a How does ultrasound do this?

_____

_____ [3 marks]

b Why must **high-powered** ultrasound be used?

_____ [2 marks]

D–C

Treatment

D–C

1 a Why are X-rays and gamma rays suitable to treat cancer patients?

_____
_____ [2 marks]

b Why are alpha and beta particles **not** suitable to treat cancer patients?

_____
_____ [2 marks]

B–A* **c** Explain how gamma radiation is emitted by the nuclei of a radioactive substance.

_____
_____ [3 marks]

D–C

2 a What is a radioactive tracer?

_____
_____ [2 marks]

b Why is it used?

_____
_____ [2 marks]

c What sort of radiation should a tracer emit?

_____ [1 mark]

d Which organ of the body is investigated using iodine-123 as a tracer?

_____ [1 mark]

e X-rays and gamma rays have similar properties. Why are gamma rays suitable to use as tracers but X-rays are not?

_____ [1 mark]

B–A*

3 Three similar sources of radiation are used to destroy a brain tumour.

a i What type of radiation would be most suitable?

_____ [1 mark]

source of radiation

tumour

source of radiation

source of radiation

ii Give a reason for your choice.

_____
_____ [2 marks]

b Why are there **three** sources of radiation arranged as shown?

_____
_____ [2 marks]

c Describe an alternative technique to destroy a brain tumour that uses only **one** source of radiation.

_____
_____ [2 marks]

What is radioactivity?

1 a Finish the table about the three types of nuclear radiation.

type of radiation	charge (+, – or 0)	what it is	particle or wave
alpha			
beta			
gamma			

[5 marks]

b Name the type of nuclear radiation that

is the most penetrating _____

is stopped by several sheets of paper _____

has the greatest mass_____

does not change the composition of the nucleus _____

travels at about one-tenth the speed of light _____

[5 marks]

2 a What is meant by 'half-life'?

_____

_____

[2 marks]

b The graph shows how the activity of cobalt-60 changes with time. Use the graph to find the half-life of cobalt. Show clearly, on the graph, how you got your answer. _____

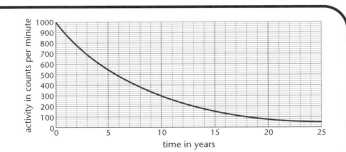

[2 marks]

3 Radon, $^{220}_{86}$Rn, is radioactive. It decays to an isotope of polonium, $^{216}_{84}$Po with a half-life of 52 s.

a How many protons are there in a radon nucleus? _____ [1 mark]

b How many neutrons are there in a radon nucleus? _____ [1 mark]

c What is the name of the particle emitted in this decay? _____ [1 mark]

d i Write a nuclear equation to describe the decay of a radon nucleus.

_____ [3 marks]

ii Show that the atomic mass and mass number are conserved in this decay.

_____ [2 marks]

e When a beta particle is emitted from a nucleus the mass number is unchanged but the atomic number increases by one. Explain how this is possible.

_____

_____ [2 marks]

f Finish the nuclear equation for the decay of iodine-131, emitting a beta particle.

$^{131}_{53}$I $\longrightarrow$ $^{x}_{y}$Xe $+ ^{0}_{-1}\beta$ x = _____

y = _____ [2 marks]

Uses of radioisotopes

D–C

1 a Suggest **two natural** sources of background radiation.

_____ [2 marks]

B–A*

b Suggest **two** sources of background radiation that arise from human activity.

_____ [2 marks]

B–A*

2 Harry works for an oil company. A leak has been reported in an underground pipe. He decides to locate the leak by introducing a small amount of radioisotope into the pipe.

a What sort of radiation should the radioisotope emit? Explain your choice.

_____

_____ [3 marks]

b What radiation detector could Harry use?

_____ [1 mark]

c How will Harry tell the site of the leak from his results?

_____ [1 mark]

D–C

3 Smoke alarms use a source of alpha radiation in a small chamber.

a Why is alpha radiation more suitable than either gamma or beta radiation for use in a smoke alarm?

_____ [1 mark]

b Explain how the smoke alarm works.

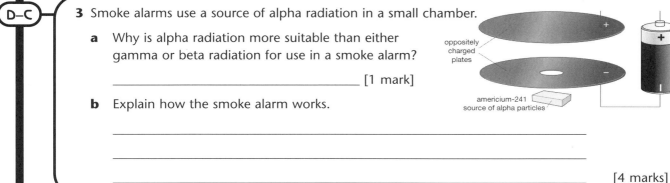

oppositely charged plates

americium-241 source of alpha particles

_____

_____

_____ [4 marks]

---

4 Trees contain carbon-14 which is radioactive. The graph shows how the activity of 1 kg of wood changes after a tree has died. This can be used to estimate the age of objects that were once alive.

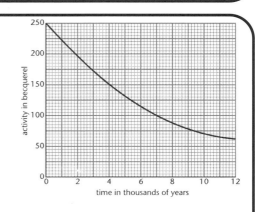

D–C

a This method cannot be used to date a wooden bowl made from a tree that died less than 200 years ago. Explain why.

_____

_____ [1 mark]

b The remains of an ancient civilisation were found. Put a ring around the objects that could be dated using this method.

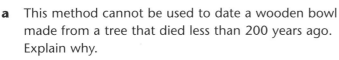

    animal bone   **bronze tool**   **seeds**   **stone**   [2 marks]

c Why can carbon-14 not be used to date rocks?

_____ [1 mark]

B–A*

d 1 kg of wood found in an archaeological dig had an activity of 150 Bq. Use the graph to estimate its age. Show your answer on the graph. [2 marks]

Fission

1 a Use these words to complete the following sentences explaining how a power station works.

generator   source of energy   steam   turbine   water

D–C

The _____ provides heat to boil the _____ to produce _____.

The pressure of the _____ turns the _____ which turns

the _____ making electricity.   [6 marks]

b What is meant by 'fission'?

_____
                                                                 [2 marks]

c Enriched uranium is used as the fuel in a nuclear power station.
What is 'enriched uranium'?

_____
                                                                 [2 marks]

d A slow neutron can be captured by a uranium-235 nucleus and the 'new' nucleus then becomes unstable.

  i Describe what happens after the capture of the neutron by the uranium nucleus.

_____

_____
                                                                 [3 marks]

  ii How can this one event become a continuous chain reaction?

_____

_____
                                                                 [2 marks]

2 Nuclear power stations produce radioactive waste.

D–C

a Suggest **one** method of disposing of low level radioactive waste.

_____
                                                                 [1 mark]

b Suggest **one** method of disposing of high level radioactive waste, such as spent fuel rods from a nuclear reactor.

_____
                                                                 [1 mark]

3 a In one type of nuclear power station graphite is used as a moderator and boron rods are used to control the number of nuclear fissions in a given time.

B–A*

  i What does a 'moderator' do?

_____
                                                                 [1 mark]

  ii Why is a moderator necessary?

_____
                                                                 [1 mark]

  iii How do the boron rods control the number of fissions?

_____

_____
                                                                 [2 marks]

b State **one** advantage and **one** disadvantage of using nuclear energy to generate electricity compared with using coal as a fuel.

Advantage _____

Disadvantage _____
                                                                 [2 marks]

P4 Revision checklist

- I can describe static electricity in terms of the movement of electrons. ☐

- I can explain how static electricity can be dangerous and how the risk of shock can be reduced. ☐

- I can describe and explain some uses of static electricity. ☐

- I can explain the behaviour of simple circuits and how resistors are used in circuits. ☐

- I can state and use the formula: resistance = voltage ÷ current, including a change of subject. ☐

- I can explain the function of live, neutral and earth wires, circuit breakers and double insulation. ☐

- I can explain how a wire fuse protects an appliance. ☐

- I can describe the motion of particles in longitudinal and transverse waves. ☐

- I can explain how ultrasound is used in body scans and to break down kidney stones. ☐

- I can explain how nuclear radiation is used in hospitals to treat cancer and as a tracer. ☐

- I can describe radioactive decay and explain and use the concept of half-life. ☐

- I can describe alpha and beta decay and construct nuclear equations to explain it. ☐

- I can explain the origins of background radiation and some non-medical uses of radioisotopes. ☐

- I can describe in detail how domestic electricity is generated in a nuclear power station. ☐

The periodic table

Key

relative atomic mass	1
atomic symbol	**H**
name	hydrogen
atomic (proton) number	1

1	2												3	4	5	6	7	8
																		4 **He** helium 2
7 **Li** lithium 3	9 **Be** beryllium 4												11 **B** boron 5	12 **C** carbon 6	14 **N** nitrogen 7	16 **O** oxygen 8	19 **F** fluorine 9	20 **Ne** neon 10
23 **Na** sodium 11	24 **Mg** magnesium 12												27 **Al** aluminium 13	28 **Si** silicon 14	31 **P** phosphorus 15	32 **S** sulfur 16	35.5 **Cl** chlorine 17	40 **Ar** argon 18
39 **K** potassium 19	40 **Ca** calcium 20	45 **Sc** scandium 21	48 **Ti** titanium 22	51 **V** vanadium 23	52 **Cr** chromium 24	55 **Mn** manganese 25	56 **Fe** iron 26	59 **Co** cobalt 27	59 **Ni** nickel 28	63.5 **Cu** copper 29	65 **Zn** zinc 30		70 **Ga** gallium 31	73 **Ge** germanium 32	75 **As** arsenic 33	79 **Se** selenium 34	80 **Br** bromine 35	84 **Kr** krypton 36
85 **Rb** rubidium 37	88 **Sr** strontium 38	89 **Y** yttrium 39	91 **Zr** zirconium 40	93 **Nb** niobium 41	96 **Mo** molybdenum 42	[98] **Tc** technetium 43	101 **Ru** ruthenium 44	103 **Rh** rhodium 45	106 **Pd** palladium 46	108 **Ag** silver 47	112 **Cd** cadmium 48		115 **In** indium 49	119 **Sn** tin 50	122 **Sb** antimony 51	128 **Te** tellurium 52	127 **I** iodine 53	131 **Xe** xenon 54
133 **Cs** caesium 55	137 **Ba** barium 56	139 **La*** lanthanum 57	178 **Hf** hafnium 72	181 **Ta** tantalum 73	184 **W** tungsten 74	186 **Re** rhenium 75	190 **Os** osmium 76	192 **Ir** iridium 77	195 **Pt** platinum 78	197 **Au** gold 79	201 **Hg** mercury 80		204 **Tl** thallium 81	207 **Pb** lead 82	209 **Bi** bismuth 83	[209] **Po** polonium 84	[210] **At** astatine 85	[222] **Rn** radon 86
[223] **Fr** francium 87	[226] **Ra** radium 88	[227] **Ac*** actinium 89	[261] **Rf** rutherfordium 104	[262] **Db** dubnium 105	[266] **Sg** seaborgium 106	[264] **Bh** bohrium 107	[277] **Hs** hassium 108	[268] **Mt** meitnerium 109	[271] **Ds** darmstadtium 110	[272] **Rg** roentgenium 111								

Elements with atomic numbers 112–116 have been reported but not fully authenticated.

* The Lanthanides (atomic numbers 58–71) and the Actinides (atomic numbers 90–103) have been omitted.
Cu and Cl have not been rounded to the nearest whole number.